AF595329

Interactions of Positrons with Matter and Radiation

Interactions of Positrons with Matter and Radiation

Editor

Anand K. Bhatia

MDPI • Basel • Beijing • Wuhan • Barcelona • Belgrade • Manchester • Tokyo • Cluj • Tianjin

Editor
Anand K. Bhatia
Heliophysics Science Division,
NASA/Goddard Space Flight Center
USA

Editorial Office
MDPI
St. Alban-Anlage 66
4052 Basel, Switzerland

This is a reprint of articles from the Special Issue published online in the open access journal *Atoms* (ISSN 2218-2004) (available at: https://www.mdpi.com/journal/atoms/special_issues/Positron-phys).

For citation purposes, cite each article independently as indicated on the article page online and as indicated below:

LastName, A.A.; LastName, B.B.; LastName, C.C. Article Title. *Journal Name* **Year**, *Volume Number*, Page Range.

ISBN 978-3-03943-795-5 (Hbk)
ISBN 978-3-03943-796-2 (PDF)

Contents

About the Editor

Anand K. Bhatia is Scientist Emeritus at NASA/Goddard Space Flight Center in Greenbelt, Maryland, USA. Before receiving his PhD from the University of Maryland in 1963, he taught for a year at the Wesleyan University in Connecticut. He was a National Academy Research Fellow from 1963 to 1965 and then joined the staff at Goddard. He developed the method of symmetric three-body Euler angle decomposition (with A. Temkin). He specializes in calculations of bound states, Feshbach resonances, and scattering of electrons and positrons using his hybrid theory. He has applied his results to the opacity of the atmosphere of the Sun and the opacity of the late-type stellar atmosphere, pointing out the importance of positrons in addition to electrons. He is a fellow of the American Physical Society.

Preface to "Interactions of Positrons with Matter and Radiation"

Dirac, in 1928, predicted the antiparticle of the electron. Positrons, produced by cosmic rays in a cloud chamber, were detected by Anderson in 1932. Since then, positron interactions, like electron interactions, with matter and radiation, have been studied extensively, both theoretically and experimentally. Theoretical calculations could have been easier because of the absence of exchange but positronium formation has to be considered in most processes. Positrons have been useful in hospitals. Positron emission tomography (PET) scans are used in hospitals to diagnose metabolic activity in the human body. Positrons and electrons can form positronium atoms, which annihilate, giving 0.511-MeV gamma rays. Such rays have been observed from the center of the galaxy. Positron annihilation has been used to detect defects in metals. In this Special Issue, we collected publications on scattering, excitation, resonances, threshold laws, and the formation of antihydrogen, which can be used to study whether the laws of quantum mechanics are the same for matter and antimatter.

Anand K. Bhatia

Editor

Article

A Note on the Opacity of the Sun's Atmosphere

Anand. K. Bhatia * and William. D. Pesnell

Heliophysics Science Division, NASA/Goddard Space Flight Center, Greenbelt, MD 20771, USA; william.d.pesnell@nasa.gov

* Correspondence: anand.k.bhatia@nasa.gov

Received: 12 June 2020; Accepted: 15 July 2020; Published: 21 July 2020

Abstract: The opacity of the atmosphere of the Sun is due to processes such as Thomson scattering, bound–bound transitions and photodetachment (bound–free) of hydrogen and positronium ions. The well-studied free–free transitions involving photons, electrons, and hydrogen atoms are re-examined, while free–free transitions involving positrons are considered for the first time. Cross sections, averaged over a Maxwellian velocity distribution, involving positrons are comparable to those involving electrons. This indicates that positrons do contribute to the opacity of the atmosphere of the Sun. Accurate results are obtained because definitive phase shifts are known for electron–hydrogen and positron–hydrogen scattering.

Keywords: photodetachment; free–free transitions; opacity

---

1. Introduction

The variation of the solar spectral irradiance with wavelength shows the effects of bound–bound, bound–free, and free–free opacity of many elements in the solar atmosphere. In 1923, using a classical approach, Kramers [1] showed that the free–free absorption coefficient is given by

$$k_{\nu} = Z^2 \rho T^{-1/2} \nu^{-3} \tag{1}$$

The solar medium is opaque between 4000 and 25000 Å because of various processes, such as Thomson scattering, bound–bound transitions, photodetachment (bound–free) or free–free transitions. In 1939, Wildt [2] suggested that an important source of opacity in the solar atmosphere could be due to the photodetachment of negative hydrogen ions:

$$h\nu + H^- \rightarrow e + H \tag{2}$$

The bound–free transitions explain the opaqueness of the Sun's atmosphere between 4000 and 16,000 Å; beyond this range, free–free transitions account for the continuous spectrum of the Sun

$$h\nu + e^- + H \rightarrow e^- + H \tag{3}$$

If an electron with energy k_0^2 absorbs photon energy and the final energy of the electron in the continuum is k_1^2, the change in energy is $\Delta k^2 = |k_0^2 - k_1^2|$. These transitions also explain the opacity of late-type stellar atmospheres. Cross sections for bound–free and free–free transitions have been calculated by Chandrasekhar and Elbert [3] and Chandrasekhar and Breen [4], respectively. The cross section for free–free transition, given by the latter [4] is

$$\sigma(k_0^2, \Delta k^2) = \frac{256\pi^2}{3} (\frac{2\pi e^2}{hc}) (\frac{h^2}{4\pi^2 m e^2})^5 \frac{1}{k_0^2 k_1 (\Delta k^2)^3} M(k_0, k_1) \text{ cm}^2 \tag{4}$$

where various quantities have the usual meaning and

$$M(k,k_1) = \left|M(0,k^2|1,{k_1}^2)\right|^2 + \left|M(0,{k_1}^2|1,k^2)\right|^2 \tag{5}$$

$$M(0,k^2|1,k_1^2) = \frac{k_1^4}{16}(3\ \sin^2\delta_k^- + \sin^2\delta_k^+) \tag{6}$$

The phase shifts δ_k^- and δ_k^+ are the triplet and singlet phase shifts for the scattering of an electron from a hydrogen atom with momentum k. They were calculated using hybrid theory [5]. The present electron–hydrogen phase shifts are much more accurate compared to those used in earlier calculations for calculating cross sections for free–free transitions. The hybrid theory takes into account exchange, short-range correlations, and long-range correlations, at the same time. There are a number of earlier calculations. For example, the calculations in ref. [6] include only long-range correlations, while in ref. [7], only short-range correlations could be considered.

It has been known that there are positrons present in the Sun [8] and in interstellar space, as indicated by the detection of the 0.511 MeV line from the center of the galaxy due to the annihilation of the positron and electron pairs [9,10]. Positrons are produced due to various processes: when two protons collide, during the formation of ^{3}He nuclei, the decay of radioactive nuclei, and the decay of positive pions to muons, which further decay into positrons [11]. Positrons produced by solar flares can reach the solar atmosphere and modulate the radiant flux passing through them during their lifetime.

Once positrons are available, the photodetachment of negative positronium ions

$$h\nu + \mathrm{Ps}^- \rightarrow e^- + \mathrm{Ps} \tag{7}$$

and free–free transitions of positrons on H are possible:

$$h\nu + e^+ + H \rightarrow e^+ + H \tag{8}$$

For the latter, there is only one phase shift for a positron with an incident momentum k and Equation (6) takes the form

$$M(0,k^2|1,k_1^2) = \frac{k_1^4 \sin^2(\delta_k)}{4} \tag{9}$$

The positron–hydrogen phase shifts (δ_k) were calculated using hybrid theory [12]. An earlier calculation of reference [13] included only short-range correlations.

The cross section for photodetachment of H^- is given by Ohmura and Ohmura [14] as

$$\sigma(H^-) = \frac{6.8475 \times 10^{-18}\gamma k^3}{(1-\gamma\rho)(\gamma^2+k^2)^3}\mathrm{cm}^2 \tag{10}$$

where $\gamma = 0.2365833$ and $\rho = 2.646$.

Photodetachment cross sections of negative positronium ions [15] were calculated using Ps^- wave functions of the form used by Ohmura and Ohmura [14] for the negative hydrogen ion. This cross section is written in the form

$$\sigma(Ps^-) = \frac{1.32 \times 10^{-18}k^3}{(\gamma^2+k^2)^3}\mathrm{cm}^2 \tag{11}$$

where $\gamma = 0.12651775$, and because $1.5\gamma^2$ is the binding energy, this gives a value of 0.024010 Ry [15]. In the above equations, k is the momentum of the outgoing electron. A measurement of this cross section was reported in [16].

We list the cross sections for bound–free and free–free transitions for electrons and positrons in Table 1 (and show them in Figure 1), where we have assumed a temperature of 6300 K and used the H^-/H ratio given by Wheeler and Wildt [17]. This temperature has been used for many years as representative of a typical cool star, including the Sun at an optical depth of approximately 1. We see that the contribution of positrons cannot be neglected in calculations of the opacity of the Sun's atmosphere.

Table 1. Comparison of bound–free (σ_{bf}) and free–free (σ_{ff}) cross sections (cm^2) for electrons and positrons, $T = 6300$ K.

		Electrons			Positrons		
Δk^2	λ (Å)	σ_{bf} *	σ_{ff}	$\sigma_{bf} + \sigma_{ff}$	σ_{bf}	σ_{ff}	$\sigma_{bf} + \sigma_{ff}$
0.26	3505	2.29 (−17)	4.28 (−20)	2.29 (−17)	9.95 (−18)	4.14 (−21)	9.95 (−18)
0.12	7594	4.15 (−17)	1.88 (−19)	4.17 (−17)	3.17 (−17)	1.56 (−20)	3.17 (−17)
0.10	9113	4.13 (−17)	2.69 (−19)	4.16 (−17)	4.17 (−17)	2.15 (−20)	4.17 (−17)
0.06	15,188	7.05 (−18)	7.45 (−19)	7.80 (−18)	8.96 (−17)	5.38 (−20)	8.97 (−17)
0.04	22,783	0.00	1.68 (−18)	1.68 (−18)	1.65 (−16)	1.13 (−19)	1.65 (−16)
0.03	30,377	0.00	2.99 (−18)	2.99 (−18)	2.53 (−16)	1.96 (−19)	2.53 (−16)
0.02	45,565	0.00	6.74 (−18)	6.74 (−18)	4.64 (−16)	4.30 (−19)	4.64 (−16)
0.01	91,130	0.00	2.70 (−17)	2.70 (−17)	1.30 (−15)	1.68 (−18)	1.30 (−15)
0.005	182,260	0.00	1.08 (−16)	1.08 (−16)	3.63 (−15)	6.72 (−18)	3.64 (−15)
0.003	303,767	0.00	3.00 (−16)	3.00 (−16)	7.69 (−15)	1.87 (−17)	7.71 (−15)
0.001	911,300	0.00	2.70 (−15)	2.70 (−15)	3.55 (−14)	1.68 (−16)	3.57 (−14)

* The number in parentheses indicates the power of ten multiplying that entry.

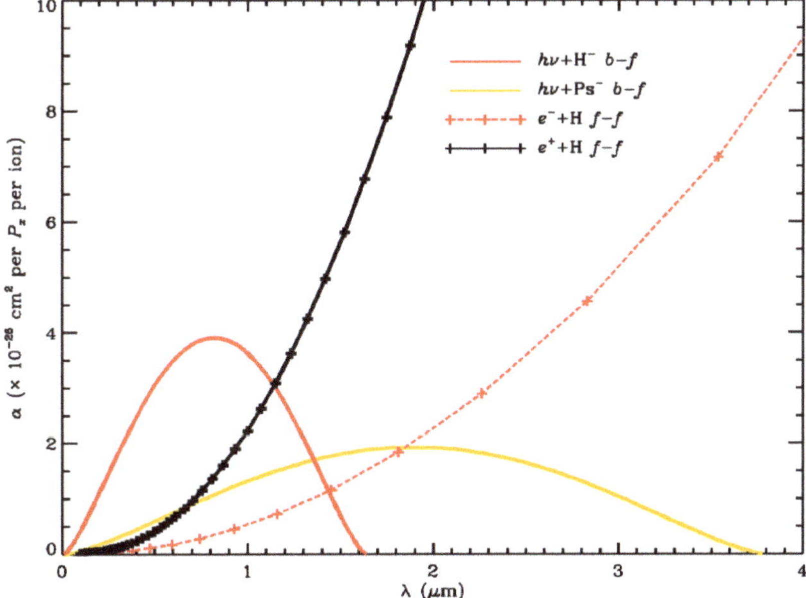

Figure 1. Absorption coefficients of four scattering processes with T = 6300 K ($\theta = 0.8$). The H^- bound–free transitions are described by Equation (10) and the Ps^- b–f transitions by Equation (11). The b–f absorption coefficients are formed by multiplying the cross sections by either the fraction of H^-/H or Ps^-/Ps, given by the Saha equation. The free–free transitions are for electrons on H (Equations (4) and (6)) and positrons on H (Equations (4) and (9)). The H b–f and f–f absorption coefficients are in units of 10^{-26} cm^2 per P_e per H^- atom, where P_e is the electron pressure. The Ps^- bound–free coefficients are in units of 10^{-26} cm^2 per P_e per Ps^- atom. The positron–H free–free coefficients are in units of 10^{-26} cm^2 per P_{e+} per H^- atom, where P_{e+} is the thermal pressure of positrons.

Our results are shown in Figure 1. An examination of Figure 1 shows several effects. The most noticeable effect of positrons on the emergent spectrum would be an increase in the brightness temperature of wavelengths longer than 1 μm. The positron–H free–free opacity is larger than the electron–H free–free opacity at those wavelengths. The second effect is when the Ps abundance becomes appreciable and a broad region of opacity appears between 0.1 and 4 μm. A third effect, positron–Ps free–free transitions, will be described in a future work.

2. Conclusions

In addition to bound–free transitions, free–free transitions are important in the solar as well as stellar atmospheres. We have calculated cross sections for these processes and have shown that the free–free transitions involving electrons dominate at wavelengths longer than 16,000 Å. The same processes are present when positrons are involved in transitions instead of electrons. Two observable quantities, the locations of the maximum in the bound–free opacity and the transition from dominance of bound–free to free–free opacity, are both located at longer wavelengths when positrons and the negative positronium ions are considered. The presence of these shifted features would be a unique marker of an object with a measurable number of positrons in its atmosphere. Processes involving positrons cannot be neglected and their contribution to opacity could be comparable to those involving electrons.

Author Contributions: A.K.B. and W.D.P. contributed equally. All authors have read and agreed to the published version of the manuscript.

Funding: No funding was received for this research.

Conflicts of Interest: The authors declare no conflict of interest.

References

1. Kramers, H.A. On the theory of X-ray absorption and the continuous X-ray spectrum. *Philos. Mag.* **1923**, *46*, 836–871. [CrossRef]
2. Wildt, R. Electron affinity in astrophysics. *Astrophys. J.* **1939**, *89*, 295–301. [CrossRef]
3. Chandrasekhar, S.; Elbert, D.D. On the continuous absorption coefficient of the negative hydrogen ion. IV. *Astrophys. J.* **1958**, *128*, 633. [CrossRef]
4. Chandrasekhar, S.; Breen, F.H. On the continuous absorption coefficient of the negative hydrogen ion. *Astrophys. J.* **1946**, *104*, 430. [CrossRef]
5. Bhatia, A.K. Hybrid theory of electron-hydrogen scattering. *Phys. Rev. A* **2007**, *75*, 032713. [CrossRef]
6. Temkin, A.; Lamkin, J.C. Application of the Method of Polarized Orbitals to the Scattering of Electrons from Hydrogen. *Phys. Rev.* **1961**, *121*, 788. [CrossRef]
7. Bhatia, A.K.; Temkin, A. Complex correlation Kohn-T method of calculating total and elastic cross sections: Electron-Hydrogen elastic scattering. *Phys. Rev. A* **2001**, *64*, 032709. [CrossRef]
8. Gopalswamy, N. Positron processes in the Sun. *Atoms* **2020**, *8*, 14. [CrossRef]
9. Knödlseder, J.; Jean, P.; Lonjou, V.; Weidenspointner, G.; Guessoum, N.; Gillard, W.; Skinner, G.; von Ballmoos, P.; Vedrenne, G.; Roques, J.-P.; et al. The all-sky distribution of 511 keV electron-positron annihilation emission. *Astron. Astrophys.* **2005**, *441*, 513–532. [CrossRef]
10. Leventhal, M. Recent Balloon Observation of the Galactic Center 511 keV annihilation line. *Adv. Space Res.* **1991**, *11*, 157. [CrossRef]
11. Murphy, R.J.; Share, G.H.; Skibo, J.G.; Kozlovsky, B. The physics of positron annihilation in the solar atmosphere. *Astrophys. J. Suppl. Ser.* **2005**, *161*, 495. [CrossRef]
12. Bhatia, A.K. Positron-Hydrogen scattering, annihilation and positronium formation. *Atoms* **2016**, *4*, 27. [CrossRef]
13. Bhatia, A.K.; Temkin, A.; Drachman, R.J.; Eiserike, H. Generalized Hylleraas calculations of Positron-Hydrogen Scattering. *Phys. Rev. A* **1971**, *3*, 1328. [CrossRef]
14. Ohmura, T.; Ohmura, H. Electron-Hydrogen scattering at low energies. *Phys. Rev.* **1960**, *118*, 154. [CrossRef]

15. Bhatia, A.K.; Drachman, R.J. Photodetachment of positronium negative ion. *Phys. Rev. A* **1985**, *32*, 1745. [CrossRef] [PubMed]
16. Michishio, K.; Tachibana, T.; Terabe, H.; Igarashi, A.; Wada, K.; Kuga, T.; Yagishita, A.; Hyodo, T.; Nagashima, Y. Photodetachment of positronium negative ions. *Phys. Rev. Lett.* **2011**, *106*, 153401. [CrossRef] [PubMed]
17. Wheeler, J.A.; Wildt, R. The absorption coefficient of the free-free transitions of the negative Hydrogen ion. *Astrophys. J.* **1942**, *95*, 281. [CrossRef]

Article

Nature's Pick-Up Tool, the Stark Effect Induced Gailitis Resonances and Applications

Chi-Yu Hu [1,*] and David Caballero [2]

1 Department of Physics and Astronomy, California State University, Long Beach, CA 90840, USA
2 The Boeing Company, Huntington Beach, CA 92647, USA; Dcaballero@socal.rr.com
* Correspondence: Chiyu.Hu@CSULB.edu

Received: 31 March 2020; Accepted: 15 June 2020; Published: 2 July 2020

Abstract: A simple universal physical mechanism hidden for more than half a century is unexpectedly discovered from a calculation of low excitation antihydrogen. For ease of reference, this mechanism is named Gailitis resonance. We demonstrate, in great detail, that Gailitis resonances are capable of explaining p+^{7}Li low energy nuclear fusion, *d-d* fusion on a *Pd* lattice and the initial transient fusion peak in muon catalyzed fusion. Hopefully, these examples will help to identify Gailitis resonances in other systems.

Keywords: Stark effects; Gailitis resonance; LENR; muon catalyzed fusion

---

1. Uncovering the Truth from the Nature Takes Time

During the 1960's, Gailitis and Damburg [1] found weak oscillations in the scattering matrix at energy ε slightly above $H(n = 2)$ energy level in electron-hydrogen scattering calculations. They suggested that the oscillations might have originated from the electric dipole component of the target system.

In an attempt to help the antihydrogen research, a calculation of the total cross section for antihydrogen formation for $\overline{H}(n \leq 2)$ using nine partial waves was carried out [2] for the following reaction, using the modified Faddeev equation,

$$\overline{p} + Ps(n = 2) \rightarrow e + \overline{H}(n \leq 2) \tag{1}$$

Near the energy region of the Gailitis oscillation, a very large $\overline{H}(n \leq 2)$ formation cross section of $1397\pi a_0{}^2$ was found, where a_0 is the Bohr radius. The S-partial wave portion of reaction (1) is plotted in Figure 1 in the energy range between the $Ps(n = 2)$ threshold to the $\overline{H}(n = 3)$ threshold.

A couple of resonant-like peaks are clearly visible. However, no previous calculations indicated any resonance in this energy region and subsequent independent calculations also could not reproduce it. Their calculations were carried out with a much shorter cutoff radius than the 450 Bohr radius used in Figure 1.

Ten years later, we decided to investigate the energy region where the questionable cross section peaks were found in Figure 1, using a much larger size job allowed by more powerful computers. The cutoff radius used in [3,4] is 1000 Bohr radius for the following S-state and with six open channel charge conjugation system.

$$\begin{array}{l} \text{Channel\#IncomingChannel} \\ 1.e^{+} + H(n=1) \\ 2.e^{+} + H(n=2,\, l=0) \\ 3.e^{+} + H(n=2,\, l=1) \\ 4.p + Ps(n=1) \\ 5.p + Ps(n=2,\, l=0) \\ 6.p + Ps(n=2,\, l=1) \end{array} \tag{2}$$

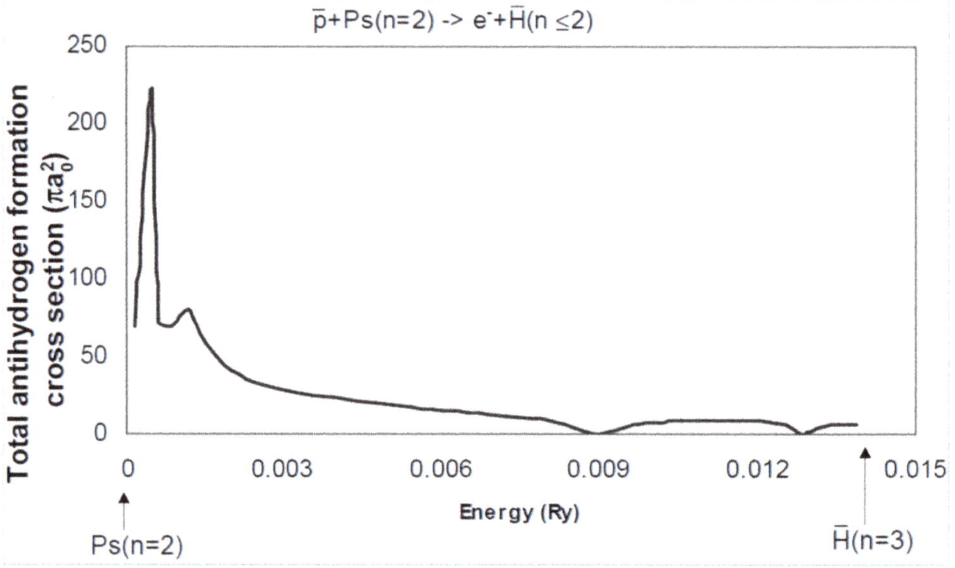

Figure 1. Total S-state antihydrogen formation cross section. Taken from reference [2]. The relatively large cutoff radius of 450 a_0 used enabled two Gailitis resonances appear near the threshold of $Ps(n = 2)$.

After numerically solving nearly half a million coupled linear equations, complete sets of beautiful 6×6 scattering matrices near each of the three resonances were obtained [3,4].

Resonances occur only in channels 5 and 6 of Equation (2), due to their large electric dipole moment. Near resonances, the Faddeev wave amplitudes of channels 5 and 6 and their scattering matrices, $tan(\delta_{ii})$, $i = 5,6$, are presented in Figure 2 below.

Apparently, the largest possible run with $y_{max} \approx 1000a_0$ used in our calculations is too short, the third resonance get cut in half. In spite of such defects, these graphs provide enough information to reveal real physics.

Figure 2a is a plot of the K-matrix elements $tan(\delta_{ii})$, i = 5,6, as a function of the energy E_1, the collision energy with respect to channel 1. Other channel energies are determined in term of E_1. For example the channel energy for channel ε_5 and ε_6 are measured from $Ps(n = 2)$, while E_1 measured from $H(n = 1)$.

Figure 2b is a plot of the Faddeev wave amplitude for channel 6 at energy close to the third resonance.

Figure 2c is a plot of the Faddeev wave amplitude for channel 5 at energy near the second resonance.

Figure 2d is a plot of the Faddeev wave amplitude for channel 5 at energy, not close enough to the first resonance.

All the wave amplitudes of the other Faddeev channels are orders of magnitude smaller.

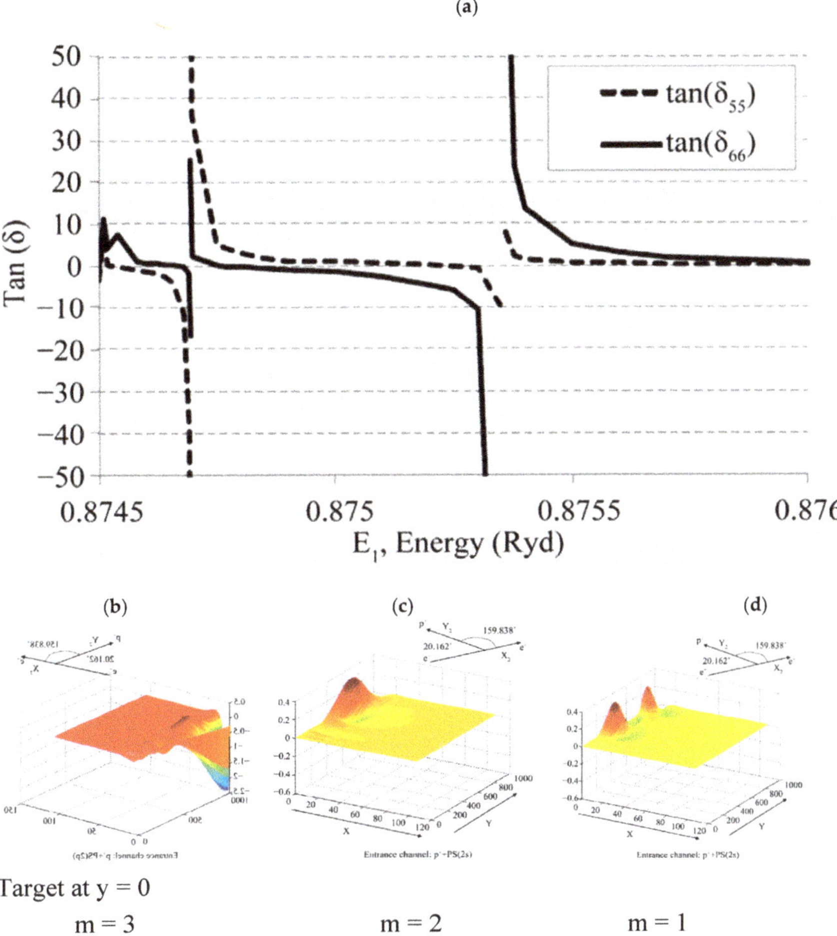

Figure 2. **(a–d)** are taken from reference [3]. The only 6 open Channel modified Faddeev Equation Calculation to date. More details can be found in Reference [3,4]. The cutoff radius used is 1000 a_0.

When the scattering matrix shows singular behavior like Figure 2a, the traditional methods that search for poles in the complex energy plane measure energy and width of the resonances. Here, for the first time, we demonstrate that the Faddeev amplitudes contain all the physics that can be revealed with much less effort.

2. Physics Revealed from Figure 2a

As the proton moves in the attractive electric dipole field from $Ps(n = 2)$ along the y-axis, the phase shift δ suddenly drops from 0^- to $-\pi/2$. That means that the attractive electric dipole field from $Ps(n = 2)$ suddenly turned strongly repulsive when the energy of the proton matched the electric dipole flipping energies, thus forcing the proton to give up all its energy, and it then turns into an expanding wave packet centered on y_m, where m is a quantum number. From Figure 2b–d, y_1, y_2 can be measured directly, but not y_3. The proton stripped off its energy and turns onto an expanding wave packet. That represents the first of two stages of the lifetime of the Gailitis resonance. The second stage begins when the phase shift suddenly drops from $\pi/2$ to 0^+, indicating a strong attractive force from the target. What will happen during the second stage, which depends only on the host system, is here the *Ps*. It produces the resonant reaction represented by $p + Ps(n = 2) \rightarrow e^+ + H(n \leq 2)$. Thus, Figure 2a reveals the pick-up action of the Gailitis resonances.

3. Other Common Property of the Gailitis Resonances

It is noticed that the widths of the initial wave packets are equal to their De Broglie wavelengths. In this section, we only use units such that $h = 1$. Thus, the momentum p is related to the De Broglie wavelength $p = 1/\lambda$, and $\varepsilon = p^2 = (1/\lambda)^2$. Then, the uncertainty principle is given by $\Delta y\ \Delta p = 1$. Applied to the Gailitis resonance when $\Delta p \rightarrow p$, we find $\Delta y \approx 1/p = \lambda$ and $\Delta\varepsilon = (\Delta p)^2 = p^2$.

Thus, the Gailitis resonances have another unique property, namely, $\varepsilon_m/\Delta\varepsilon_m = 1$ for all m. Due to the incoherent use of units in some of our previous calculations [5], this formula was listed incorrectly as $\varepsilon_m/\Delta\varepsilon_m = 4\pi^2$.

The lifetime of Gailitis resonances can be calculated using the uncertainty principle in the following form:

$$\varepsilon_m = (1/\Delta\varepsilon_m) \times 2.42 \times 10^{-17}\ \text{sec} \tag{3}$$

where the energy $\Delta\varepsilon_m$ is given in Rydberg.

4. Resonant Condition Read from Faddeev Amplitudes

We can relate the known electric dipole moment $|\mu_1|$ of the target $Ps(n = 2)$ in the mass normalized Jacobi coordinate system in Figure 2b–d to the resonance energies. The subscript 1 represents the electric dipole moment only, where $y^2{}_m = \langle y^2{}_m \rangle$ are directly measured at the peaks of the wave packets in Figure 2c,d. However, due to the energy we used to calculate, 2d is too far from the resonant energy. There are two peaks in the wave packet. It is easy to identify which one produced the $m = 1$ resonance (see [3,4]). Numerically, the resonant equation below comes directly from the wave functions 2c and 2d

$$\varepsilon_m = m|\mu_1|/y^2{}_m, \text{ where } m = 1, 2. \tag{4}$$

It is noticed that this is the dipole flipping condition. The value of y_3 cannot be measured from Figure 2b with our limited computer resources. Using a numerical extension of Figure 2b beyond $1000a_0$, along with help from Equation (4), y_m, for $m = 3$ was determined. The properties of the three resonances are listed in Table 1. All quantities are in mass normalized Jacobi coordinates and $|\mu_1| = 47.94778$.

Table 1. ε_m are the resonant energies in units Ryd. $\varepsilon_m/\Delta\varepsilon_m$ are determined in Section 3. λ_m and y_m are measured directly from Figure 2b–d see Reference. [3] for more detail.

m	ε_m (Ryd)	$\varepsilon_m/\Delta\varepsilon_m$	$\lambda_m(a_0)$	$y_m(a_0)$
1	5.4436(−4)	1	380.85	296.8
2	0.19436(−3)	1	637.35	702.4
3	0.84344(−4)	1	967.54	1306.0

Please notice, the ε_m resonant energy measured just above the $Ps(n = 2)$ threshold is very small. They are within the range of the fine structure energy width, where the Coulomb degeneracy is removed by the small relativistic perturbation in the pure Coulomb force Hamiltonian that split the Coulomb energy level into a number of independent energy levels depends on the angular momentum quantum number ℓ. The energy width of this fine structure energy levels is very small. All ε_m must lie within this width, thus three-body scattering calculation involves 6-open channels represented in Equation (2).

It is well known in many textbooks that an incoming charge particle p or e^+ will induce a constant electric dipole moment in the target atom. Consequently, an attractive inverse square force exists that support Gailitis resonance listed in Table 1.

These numbers can also be simply reproduced by the resonant condition Equation (4):

$$\varepsilon_m = m|\mu_1|/y^2{}_m, \qquad m = 1, 2, 3.\rightarrow \tag{5}$$

The second column $\varepsilon_m/\Delta\varepsilon_m = 1$ indicates that, in the traditional measurements of energy and width of resonances, in the complex energy plane, this ratio must be close to one and all Gailitis resonances lying on this straight line [5]. The present method provides a complete set of properties for all the Gailitis resonances outlined above. These properties are very useful in the search for Gailitis resonances for more complex systems. Such searches have already solved a number of decades old outstanding problems. A few of them are outlined below.

5. Lifetime Rate of Gailitis Resonance in Muon-Catalyzed Fusion

The molecule $dt\mu$ is composed of a deuteron and a triton nucleus bound together by a negative muon. The "size" of this molecule is much smaller than a normal molecule bound by an electron. As a result, the probability for the overlap of the two nuclei wave functions is large enough to produce the nuclear reaction:

$$dt\mu \rightarrow {}^4He + n + \mu + 17.6\ \text{MeV}. \tag{6}$$

The number of neutrons, N, counted in the laboratory is given in Figure 3. The mechanism for the formation of large initial peaks was unknown since the publication of the original data [6].

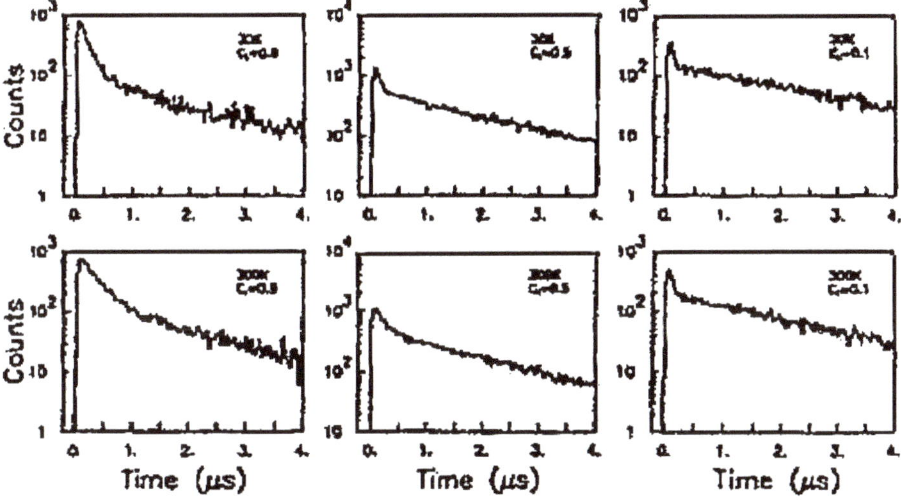

Figure 3. The initial time transient high $dt\mu$ formation peak, take from Reference [6].

In Reference [7], using the properties of Gailitis resonance and the experimental information that the initial muonic atom $t\mu$ was formed at an excited state, $n = 14$, an estimation found the lifetime of the Gailitis resonance is of the same order of magnitude as the radiative decay lifetime. This is consistent with the data shown in Figure 3. Here at the end of the life of the Gailitis resonance, as Figure 2a shows, it becomes a bound $dt\mu$ molecule before the nuclear reaction of Equation (5) takes place.

6. Long-Lived Gailitis Resonance Composed of the Electron-Rydberg Hydrogen Atom

In the Rydberg hydrogen with n~20, the electron was trapped at a very large distance, that is, a very large y_m and λ_m even for the $n = 2$ system found in Table 1. For Rydberg Gailitis resonances, the lifetime can be expected to exceed its radiation decay lifetime. That seemed to be the case for the earlier stage of research in antihydrogen production when the antihydrogen was created in Rydberg states. The expectation for such states is to radiatively cascade down to low excitation states failed.

7. Nuclear Fusion in $(p+^7Li)$ -> $^8Be^*$ -> 2α

The energy of the Gailitis resonance in $p+^7Li$ reaction overlaps with that of the compound nuclear energy level centered at 19.9 MeV, with a decay width that extends even below the proton separation energy of ^8Be (see the arrow in Figure 4).

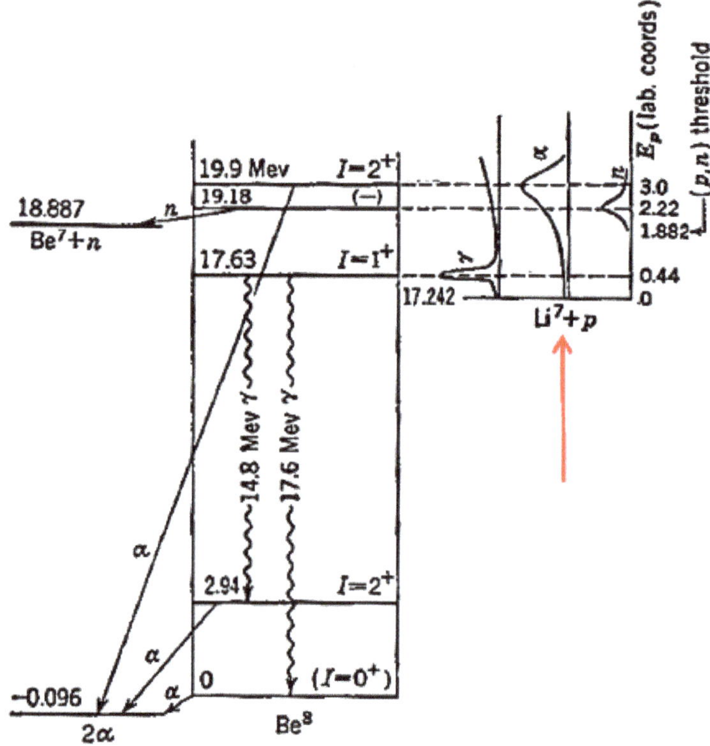

Figure 4. Some of the known energy levels of Be^8 and reactions involved in their formation and dissociation, see reference [8].

This compound nuclear energy level has $I = 2^+$, the ^7Li has $I_1 = 3/2^-$ and the proton has $I_2 = \frac{1}{2}$, with $l = 1^-$ contribution.

One possible vector sum for the resonance is $I = 2^+$, that makes a perfect match between the resonance and this compound nuclear energy level Be^* (shown by the arrow).

8. S-State Gailitis Resonance (d, Dp)—The "Quasi-Particle"

From the $e^+ + H$ calculation, it is expected that the activity of the resonance is not strong enough to disturb the lattice. In this section, we assume that the deuteron atom is bound on Pd lattice.

During the first stage of the life of (d, D_p), Figure 5 shows that the wave packet of d is a spherical layer of probability density with maximum located on a spherical surface with resonant radius r_0. A deuterium atom on the lattice is represented by D_p, and it has an electric dipole moment induced by the lattice vibrating phonon, indicating that the electric dipole can be tuned to locate the resonances. Tuning, for example, can be performed using new laser technology to induce a phonon state.

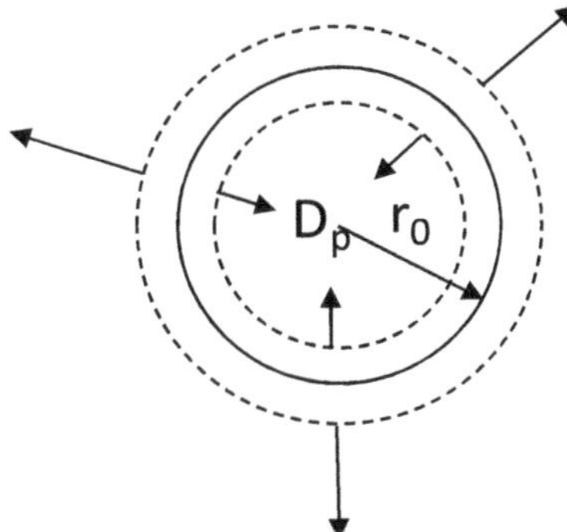

Figure 5. Cross section view of the wave-packet: an expanding spherical layer of probability density of the approaching charged particle.

The r_0 remains unchanged during the first stage of the life of the resonance, but the probability density expands both inward and outward from this surface.

During the second stage, starting with a sudden attraction towards D_p, the spherical surface with maximum probability density begins to shrink.

D_p remains on the lattice, unperturbed by the activities of the resonance. As usual, the resonant energies of Gailitis are very small for the long-lived resonances. When the probability cloud begins advancing over the Coulomb barrier, no matter how little, the quasi-particle (d, D_p) is already in a shallow negative attractive tail of the central nuclear potential, far away from the complication of the core nuclear forces.

The role of 4He+ in (d, D_p) fusion.

What does (d, D_p) have in common with $^4He^+$? They both have one electron and two deuterons d. The $^4He^+$ is the lowest possible energy system involving these three particles. The most important difference between these two systems concerning this study is the energy difference. Neglecting all small energies involving the lattice and the Gailitis resonance the (d, D_p) is 23.85 MeV above $^4He^+$. This is the energy needed to separate the two d from the α-particle, the nucleus of $^4He^+$. Unfortunately, there is no compound nuclear energy level of α to match the nuclear properties of an S-state (d, D_p). Instead of becoming a compound nuclear energy level of α at excitation energy 23.85 MeV (see Figure 6), during the second stage of the life of the (d, D_p), resonant action leads the quasi-particle (d, D_p) into a shallow negative tail of the nuclear potential of D_p, which, in an attempt to expel the intruder, injects it with energy equal to the potential energy drop, as the cloud keeps shrinking.

As soon as the energy accumulation in (d, D_p) reaches the energy of a lattice normal mode phonon, the nuclear energy begins to flow into the lattice, one phonon at a time, until the size of the "quasi-particle" shrinks close to the region where the core nuclear force dominate. Then, the nuclear force takes over the dynamics. The (d, D_p) either reaches the ground state of $^4He^+$ and still remains on the lattice, or gets kicked out of the lattice as a quasi-particle where background electrons can slow it down, until reducing it to a free $^4He^+$. This leakage of one phonon at a time is a slow process and the amount leaked each time, one phonon, is negligible for nuclear energy. Namely, only a classical conservation of energy need apply.

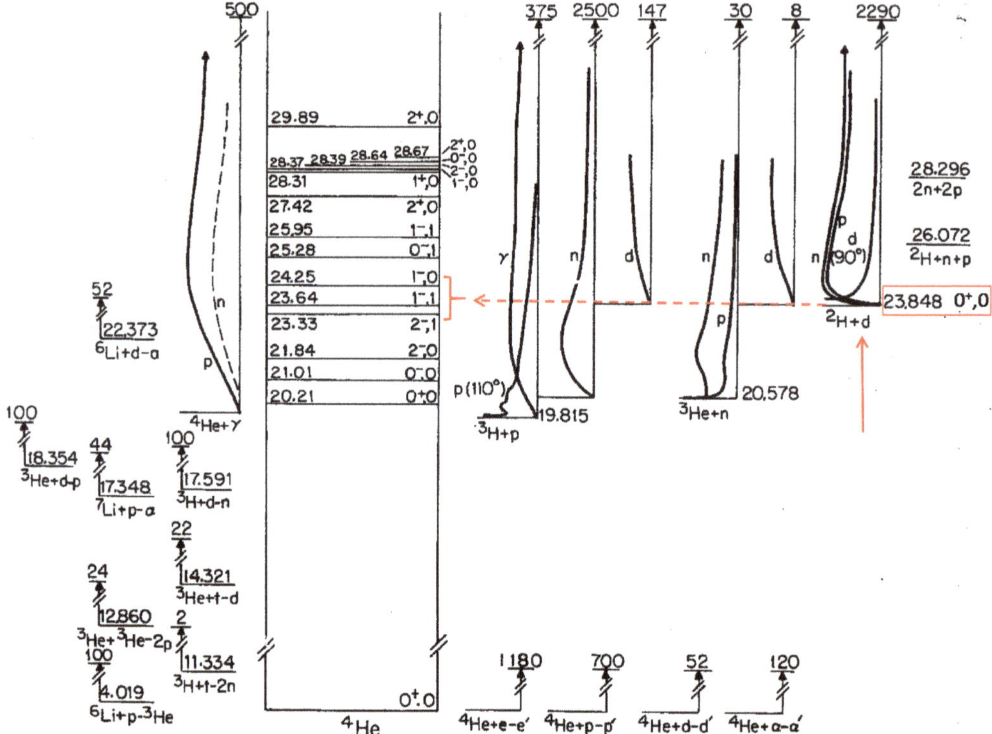

Figure 6. Energy level diagram of 4He+.

9. Discussion

The Gailitis resonances are supported by a very simple physical phenomenon. Why did it take so long to uncover its true nature? The answer is clearly displayed in Figure 2 and Table 1. The most important region is between $y = 296.8\ a_0$ and $y = 1306\ a_0$. Any computer would be hard pressed to accommodate such a large calculation correctly as seen even in some most recent calculations. In other words, the long range Coulomb force is producing unexpectedly long range physics. Present computational methods needs substantial improvements.

This problem is demonstrated clearly with our long struggle for calculating the simplest quantum three-body scattering system in $e^+ + H(n \leq 2)$, presented earlier in this report. Table 1 shows that the first Gailitis resonance shows up at $y_1 = 296.8a_0$. Had we used an effective cut-off, $y_{max} = 300a_0$, all the physics of the Gailitis resonances would have remained hidden.

Supercomputers have advanced several generations since our calculations [3,4] and the experimental techniques have also improved, so that we can make measurements in the molecular and atomic level. There is no excuse for us theorists still using the decades old methods that were designed for the needs of decades ago. It is clear that a resonance can represent its own real physical properties, which is more than just a real and imaginary part. We must start with the wave functions for each and every independent open channel.

The quantum 3-body multichannel scattering system presented in our work [3] is only the tip of the iceberg. Ref. [4] is designed to provide interested readers the first step to set up their own code. Critics should help improve it until they can find even better methods that could satisfy the needs of the 21st century.

Scientific processes brought multi-disciplinary physics closer, as shown from the small number of examples presented here; the universal natural mechanism of electric dipole flipping. A lot of other possibilities such as the role of higher order of both the electric and magnetic multipoles will be next. It seems that experimentalists are already moving ahead quickly into these areas.

Clearly, multidisciplinary collaboration can be most fruitful in such an area like the low energy nuclear fusion.

Author Contributions: C.-Y.H. did the theoretical part and the main code. D.C. helped with the graphs. All authors have read and agreed to the published version of the manuscript.

Funding: This research received no external funding.

Acknowledgments: The authors are thankful to Peter Hagelstein for the critical reading of this manuscript.

Conflicts of Interest: The author declares no conflict of interest, there might be possible miscommunication with the publishers.

References and Notes

1. Gailitis, M.; Damburg, R. The Influence of Close Coupling on the Threshold Behavior of Cross Sections of Electron-Hydrogen Scattering. *Proc. Phys. Soc.* **1963**, *82*, 192. [CrossRef]
2. Hu, C.Y.; Caballero, D.; Papp, Z. Induced Long-Range Dipole-Field-Enhanced Antihydrogen Formation in the $p + Ps(n{=}2) \rightarrow e^{+} + H(n{=}2)$ Reaction. *Phys. Rev. Letts.* **2002**, *88*, 063401. [CrossRef] [PubMed]
3. Hu, C.Y.; Caballero, D. Long-Range Correlation in Positron-Hydrogen Scattering System Near the Threshold of Ps(n = 2) Formation. *J. Mod. Phys.* **2013**, *4*, 622–627. [CrossRef]
4. Hu, C.Y. The Faddeev-Merkuriev Differential Equation (MFE) and Multichannel 3-Body Scattering Systems. *Atoms* **2016**, *4*, 16. [CrossRef]
5. Hu, C.Y.; Papp, Z. Second Order Stark-Effect Induced Gailitis Resonances in e + Ps and p + ^{7}Li. *Atoms* **2016**, *4*, 8. [CrossRef]
6. PSI data presented at the International Conference on Muon Catalyzed Fusion and Related Topics. (μCF-07) Figure 4 page 86.
7. Hu, C.Y.; Caballero, D. A Possible Role of the Gailitis Resonance in Muon Catalyzed Fusion. *J. Mod. Phys.* **2014**, *5*, 18. [CrossRef]
8. Figure 4 was taken from the Evans: *The Atomic Nucleus*; McGraw-Hill Inc.: New York NY, USA, 1955; p. 203.

Review

Positron Scattering from Atoms and Molecules

Sultana N. Nahar [1,*,†] **and Bobby Antony** [2,†]

1 Department of Astronomy, The Ohio State University, Columbus, OH 43210, USA
2 Department of Physics, Indian Institute of Technology (Indian School of Mines) Dhanbad, Jharkhand 826 004, India; bobby@iitism.ac.in
* Correspondence: nahar.1@osu.edu; Tel.: +1-614-292-1888
† These authors contributed equally to this work.

Received: 21 March 2020; Accepted: 4 June 2020; Published: 15 June 2020

Abstract: A review on the positron scattering from atoms and molecules is presented in this article. The focus on positron scattering studies is on the rise due to their presence in various fields and application of cross section data in such environments. Positron scattering is usually investigated using theoretical approaches that are similar to those for electron scattering, being its anti-particle. However, most experimental or theoretical studies are limited to the investigation of electron and positron scattering from inert gases, single electron systems and simple or symmetric molecules. Optical potential and polarized orbital approaches are the widely used methods for investigating positron scattering from atoms. Close coupling approach has also been used for scattering from atoms, but for lighter targets with low energy projectiles. The theoretical approaches have been quite successful in predicting cross sections and agree reasonably well with experimental measurements. The comparison is generally good for electrons for both elastic and inelastic scatterings cross sections, while spin polarization has been critical due to its sensitive perturbing interaction. Positron scattering cross sections show relatively less features than that of electron scattering. The features of positron impact elastic scattering have been consistent with experiment, while total cross section requires significant improvement. For scattering from molecules, utilization of both spherical complex optical potential and R-matrix methods have proved to be efficient in predicting cross sections in their respective energy ranges. The results obtained shows reasonable comparison with most of the existing data, wherever available. In the present article we illustrate these findings with a list of comprehensive references to data sources, albeit not exhaustive.

Keywords: Electron-Positron Scatterings; atoms and molecules; cross sections and spin polarization; theoretical approaches

---

1. Introduction

A positron, the antiparticle of the electron, has the same mass, electric charge (but positive) and spin (1/2) as that of an electron. Like other antiparticles, positrons were produced during the period of baryogenesis when the universe was extremely hot and dense, but now they exist in much lower numbers than its counter part, the electrons. Although not found in normal conditions, they are produced at the galatic center or supernovae events and are found in copious amount in cosmic ray showers and in the ionosphere. Positrons are created naturally in β^+ radioactive decays such as from K-40, particle reactions or by pair production from a sufficiently energetic photon interacting with the atomic nuclei in a material. Nevertheless, a small percent of potassium (0.0117%) K-40 is the single most abundant radioisotope in the human body and produces about 4000 natural positrons per day. However, soon after its creation, they annihilate with electrons or form the exotic atom, positronium (Ps), with a very short lifetime, finally decaying to 2 or 3 gamma rays each with energy 511 keV. Ps has a mass of 1.022 MeV/c^2 and can form otho-Ps (o-Ps) or a para-Ps (p-Ps) when the electron-positron spins are parallel (spin = 1) or anti-parallel (spin S = 0) respectively. An o-Ps decays to 3 and a p-Ps decays to 2 gamma photons. The energy levels of Ps are similar to those of hydrogen atom. Gamma rays, emitted indirectly by a positron-emitting radionuclide (tracer), are detected in positron emission tomography (PET) scanners used in hospitals. PET scanners create detailed three-dimensional images of metabolic activity within the human body. Positron Annihilation Spectroscopy (PAS) is used in materials research to detect variations in density, defects and displacements within a solid material. It is the detection of 511 keV gamma ray photons that is typically used as the signal for the source of existence or creation of positrons, such as, in the center of our galaxy, Milky Way.

The treatment of electron and positron scattering from atoms are similar and have been studied extensively both theoretically and experimentally. Similar is the case with molecules. The scattering parameters of interest are the scattering cross sections and the spin polarization. The scattered wave function can be obtained by solving the Schrödinger equation using Numerov method (e.g., Reference [1]), or using other methods mentioned below. The scattering parameters can be determined using the wave function. While extensive set of references are available, this review provides selected references that can lead to details of various approaches and experimental results. Most scattering studies have been carried out for neutral atoms, such as, He, Ne, Mg, Ar, Kr and Xe using various theoretical approaches, such as, polarized orbital method (e.g., References [2–5]), modified adiabatic method (e.g., Reference [6]), variational method (e.g., Reference [7]) and optical potential method (e.g., References [1,8–16]). There are many experimental studies on positron scattering as well (e.g., References [17–39]). Spin polarization for electron scattering was measured by various groups (e.g., References [40–42]). Among ions, the study has remained largely on single valence electron ions using Kohn-Feshbach variational method [43], polarized orbital method [44,45] and hybrid method [46–48]. Electron/positron scattering from molecules are investigated using spherical complex optical potential (SCOP) as well [49,50]. More references can be obtained from the cited articles. Compared to scattering from a positive ion, a neutral target offers consistent scattering features that can help in better understanding the general characteristics. Being less reactive species, noble gases are easy to handle experimentally compared other targets. They are also relatively simple collision systems to approach theoretically. Hence, it will be interesting to review the cross section data on electron/positron scattering from inert gases. The present work will concentrate on the neutral inert gases as well.

Interaction of positron with atoms is dealt mainly with two theoretical approaches; perturbative and non-perturbative. Perturbative methods usually work in the intermediate to high energy region (ionization threshold to about 10 keV), while the non-perturbative theories are capable of accurate calculations at low energies. Among various methods mentioned above, polarized orbital and optical potential methods have been used most widely to calculate scattering parameters for atoms beyond He. The polarized orbital method (e.g., Reference [51]) ansatzes the distortion in the target wave function. Temkin [2,3] first introduced it, where he included long range correlation which has the characteristic behavior of $-1/r^4$ of the longest range polarization potential, for the distortion. The method was converted to a hybrid model by Bhatia (e.g., References [48,52]), which included the short range correlation and variationally bound energies. His application of the hybrid method to calculate phase shifts, scattering lengths, photo-detachment, photoionization, positron scattering, annihilation and positronium formation produced reasonable results, which showed good agreement with available results. He extended the work to obtain accurate results in the elastic region for S-, P-, and D-wave scattering as well. Bhatia's investigation using the hybrid method focuses largely on the scattering by single-electron systems (e.g., positron impact excitation of hydrogen [45]), since the wave function of the target is known exactly. Besides, the possibility of direct annihilation and positronium formation requires a composite wavefunction, which is almost impossible to formulate.

Polarized orbital method for elastic scattering of positrons from noble gases was first developed by McEachran et al. [4] where they included in principle all multipole moments of the positron-atom interaction. The polarized orbital was calculated by a perturbed Hartree-Fock scheme, which was used to calculate the polarizability of atoms [53,54] and the scattered wave was obtained from a potential scattering problem. The method showed good agreement with measured cross sections for for He and Ne atoms [17,21]. McEachran et al. [55] implemented their method successfully for other atoms as well.

One of the most rigorous approach for positron-ion scattering has been the Feshbach projection operator method [56], where the usual Hartree-Fock and exchange potentials are augmented by an optical potential [47]. However, the method employs correlation functions that are of Hylleras type and hence do not include long range correlation.

Kohn Variational Principle (KVP) (e.g., Reference [57]) is usually applied to low-energy positron scattering to obtain elastic and Ps formation cross sections. In this method a two-component trial wave function is chosen, having the correct asymptotic form with enough flexibility to describe all the short-range distortions and correlations of the positron-atom system. This wave function is then used in the Kohn functional, which can be written in terms of the K-matrix elements. From the K-matrix, cross sections can be calculated.

The many-body-theory for a positron-atom interacting system (e.g., References [58,59]) is based on the Dyson equation. This is solved by the representation of eigenfunctions of the Hartree-Fock Hamiltonian. The formulated self-energy matrix gives the phase shifts. This approach is used to study low energy positron scattering from atoms.

Schwinger Multichannel Calculations (SMC) is a well-known method to study low energy scattering (e.g., References [60,61]). The backbone of the method is the computation of variational expression for the scattering amplitude. SMC describes target polarization through single virtual excitations of the target wave function, explicitly considered in the expansion of the scattering wave function. The Lippmann-Schwinger scattering equations are then solved to obtain the cross sections [62].

The other method of interest is the close-coupling (CC) approach for the scattered wave and the R-matrix method [63,64] to solve the coupled set of integro-differential equations. Jones et al. [26] used the convergent close-coupling (CCC) approximation, where they solve the equations with a different set of codes than standard R-matrix codes. They use multi-configuration Dirac-Fock program of Grant et al. [65] to obtain the target wave functions. Their results for positrons scattering from Ne and Ar indicate that, while both polarized orbital method and CCC approximation showed good agreement with experiment in general, the polarized orbital method yielded slightly better cross sections. In the standard close-coupling formalism for molecules (e.g., Reference [66]), the convergence in the expansion of the three-body wave function is obtained using the exact discrete eigenstates of the atomic target. The technique relies on the expansion of the total wave function in the set of target states of the atom and Ps. The CCC method allows for the examination of the effect of virtual excitation to the continuum as well [67].

For molecules, the optical potential method (e.g., References [49,68]), Born approximations (e.g., References [69,70]), and so forth are the most common quantum mechanical perturbative theories used for electron scattering presently. The positron collision studies is an extension of the optical potential method [71]. In case of non-perturbative theories, the close-coupling or the grid-based method for solving Schrödinger equation are employed. Irrespective of whether the method is perturbative or non-perturbative, the positron-molecule scattering is an extension to the positron-atom interaction technique. One has to consider multi-centre approach to deal with the projectile-target interaction due to the absence of spherical symmetry and due to the complexity of molecules. Here we will discuss few of the most commonly used theoretical methods to deal with positron-atom/molecule interaction.

Theoretical methods to investigate positron scattering and various target molecules studied by these approaches along with references are listed in Table 1. The list given below is not exhaustive, but gives an overall picture of various studies done so far. Further, this review will elaborate the most common approach, the optical potential method, to study electron and positron scattering from atoms and molecules with reasonable success.

Table 1. References for positron scattering from various target molecules using various methods: independent atom model (IAM), IAM with screening correction (IAM-SCAR) with interference (IAM-SCAR+I), Kohn variational principle (KVP), distorted wave approximation (DWA), many-body-calculation (MBC), R-matrix, Schwinger multichannel calculations (SMC), close-coupling (CC) approximation.

Method	Target(s)	Reference
IAM	H_2, NH_3, CO, CO_2, O_2, SF_6, CF_4, CCl_4, CBr_4, CI_4, CH_4, SiH_4, GeH_4, PbH_4	[72]
	H_2	[73]
	O_2, CO, CO_2, SO_2, CS_2, OCS, SF_6	[74]
	N_2	[75]
	CO_2	[76]
	Hydrocarbons	[77]
IAM-SCAR(+I)	H_2, CH_4	[78,79]
	N_2, O_2	[80]
	O_2	[81]
	H_2O	[82,83]
	N_2O	[84]
	NO_2	[85]
	Formaldehyde	[86]
	Tetrahydrofuran, 3-hydroxy-tetrahydrofuran	[87]
	Indole	[88]
	Uracil	[89]
	Pentane isomers	[90]
	2,2,4-trimethylpentane	[91]
	Vinyl acetate	[92]
	Tetrahydrofurfuryl alcohol	
	Pyrimidine	[93,94]
	Pyridine	[95]
	Cyclic ethers (oxirane, 1,4-dioxane, tetrahydropyran)	[96]
	Tetrahydrofurfuryl alcohol (THFA)	[97]
KVP	H, He	[57,98,99]
	H_2	[100]
DWA	H	
	H_2	[101]
	Inert gases	[102–106]
MBC	H	[107]
	Noble gases	[58,59]
	Mg	[108]
R-matrix	H	[63,109]
	Inert gases	[109–111]
	HF	[112]
	He_2	[113]
	H_2, N_2	[109,113–115]
	H_2O	[116]
	CO_2, Acetylene	[117,118]
SMC	He	[119,120]
	H_2	[61,120–126]
	Li_2	[127]
	N_2	[125,128–131]
	CO	[132]
	CO_2	[120,125,133,134]
	H_2O	[135–137]
	CH_4	[138]
	Formic acid	[139]
	Benzene	[140]
	Pyrimidine	[141]
	Allene	[142]

Table 1. *Cont.*

Method	Target(s)	Reference
	Silane	[62]
	THF	[143]
	Ethene	[144]
	Actetylene	[128,145]
	Ethane	[146]
	1,1-$C_2H_2F_2$	[147]
	Pyrazine	[148]
	Formaldehyde-water complexes	[149]
	C_3H_6 isomers	[150]
	Glycine and Alanine	[151]
	SF_6	[133]
	C_2H_4	[152]
	Methylamine	[153]
	Iodomethane	[154]
CC	CO	[155]
	H	[66,156–161]
	Alkali atoms	[66,162–171]
	Mg	[172,173]
	Noble gases	[174,175]
	H_2	[176–179]
	N_2	[67]

2. Scattering Parameters: Cross Sections and Spin Polarizations for Atoms

The characteristic features of the scattering can be observed in the cross section and in spin polarization caused by the projectile. While cross section can be obtained by solving non-relativistic Schrödinger or relativistic Dirac or Dirac-Fock equations, the latter provides accurate treatment for spin polarization parameters. In the present review, we will present relativistic single particle Dirac approach, which has been successful in reproducing the observed scattering phenomena.

The relativistic Dirac equation for a projection of rest mass m_o and velocity v traveling in a central field $V(r)$ is given by (e.g., Reference [180,181]),

$$[c\alpha.p + \beta m_o c^2 + V(r)]\psi = E\psi, \tag{1}$$

where α and β are the usual 4×4 Dirac matrices and ψ is a four-component (spinor) function, $\psi = (\psi_1, \psi_2, \psi_3, \psi_4)$. (ψ_1, ψ_2) are the large components and (ψ_3, ψ_4) are the small components of ψ. Defining $\gamma = (1 - v^2/c^2)^{-1/2}$, the total energy is $E = mc^2 = m_o\gamma c^2 = E' + m_o c^2$ where E' is the kinetic energy, and writing the radial function of the large large component as $G_l = \sqrt{\eta} g_l(r)/r$, the equation for the large component can be rewritten as the Dirac equation reduced to the form similar to Schrödinger equation (e.g., Reference [9]) as

$$g_l^{\pm''}(r) + \left[K^2 - \frac{l(l+1)}{r^2} - U_l^{\pm}(r)\right] g_l^{\pm}(r) = 0, \tag{2}$$

where the effective Dirac potentials due to spin up and spin down respectively are,

$$-U_l^{+}(r) = -2\gamma V + \alpha^2 V^2 - \frac{3}{4}\frac{\eta'^2}{\eta^2} + \frac{1}{2}\frac{\eta''}{\eta} + \frac{l+1}{r}\frac{\eta'}{\eta}, \tag{3}$$

and

$$-U_l^{-}(r) = -2\gamma V + \alpha^2 V^2 - \frac{3}{4}\frac{\eta'^2}{\eta^2} + \frac{1}{2}\frac{\eta''}{\eta} - \frac{l}{r}\frac{\eta'}{\eta}. \tag{4}$$

The prime and double primes represent the first- and the second-order derivatives with respect to r, $\eta = (E - V + m_o c^2)/c\hbar$, $\delta = (E - V - m_o c^2)/c\hbar$, and $K^2 = (E^2 - m_0^2 c^4)/c^2\hbar^2$. In atomic unit, $m_o = e = \hbar = 1$, $1/c = \alpha$, where α is the fine structure constant and hence $\gamma = (1 + \alpha^2 K^2)^{1/2}$ and $E = \gamma c^2 = \gamma/\alpha^2$. The proper solution of the Schrödinger like Dirac equation behaves asymptotically as,

$$g_l^{\pm''}(K,r) \sim Kr[j_l(K,r) - tan\delta_i^{\pm} n_l(K,r)], \tag{5}$$

where j_l and n_l are spherical Bessel functions of the first and second kind respectively, and $\delta_i^{\pm}$ are the phase shifts due to collisional interactions. The plus sign corresponds to the incident particles with spin up and the minus sign to those with spin downs. $\delta_i^{\pm}$ indicates the shifts in the phase of the radial wave function due to the effect of interaction potentials in the scattering. The radial wave function will be "pushed out" if the potential is repulsive and vice-versa with respect to the incoming free radial wave. So from this quantity, we can determine various microscopic quantities like cross section. The values of $\delta_i^{\pm}$ may be obtained from the values of $g_l^{\pm}$ at two adjacent points r and $r + h$ ($h << r$) as

$$\tan\delta_i^{\pm} = \frac{(r+h)g_l^{\pm}(r)j_l(K(r+h)) - rg_l^{\pm|}(r+h)j_l(Kr)}{rg_l^{\pm}(r+h)n_l(Kr) - (r+h)g_l^{\pm}(r)n_l(K(r+h))}. \tag{6}$$

The wave functions $g_l^{\pm}$ can be obtained by numerical integration of $g_l^{\pm''}$ using Numerov method and the spherical Bessel functions as described in Nahar and Wadehra [1]. Schrödinger/Dirac equation, can be solved by various other approaches mentioned above, such as, Kohn-Feshbach variational method [43], polarized orbitals method [2–4,46,47,51], close-coupling approximation [26,64] for the scattered wave function from which the phase shift is determined.

The generalized scattering amplitude for the collision process is given by [9],

$$A = f(K,\theta) + g(K,\theta)\boldsymbol{\sigma}.\hat{n}. \tag{7}$$

where

$$f(K,r) = \frac{1}{2iK}\sum_{i=0}^{\infty}\{(l+1\}[exp(2i\delta_l^{+}) - 1] + l[exp(2i\delta_l^{-}) - 1]P_l(cos\theta), \tag{8}$$

$$g(K,r) = \frac{1}{2iK}\sum_{i=0}^{\infty}[exp(2i\delta_l^{+}) - exp(2i\delta_l^{-})]P_l(cos\theta), \tag{9}$$

and $\hat{\mathbf{n}}$ is the unit vector perpendicular to the scattering plane. The differential cross section (DCS) for the scattering of the spin-1/2 particles by the spin zero neutral atom is given by

$$[DCS] = \frac{d\sigma}{d\Omega} = \sum_{\nu'} | < \chi_{\nu'}|A|\chi_\nu > |^2 = |f|^2 + |g|^2 + (f^*g + fg^*)\hat{\mathbf{n}}.\mathbf{P}_i, \tag{10}$$

where $\chi_{\nu'}$ represents a spin state and $\mathbf{P}_i =< \chi_{\nu'}|\sigma|\chi_\nu >$ is the incident-beam polarization, which is assumed to be zero. The integrated elastic cross section for the unpolarized incident beam can be obtained as

$$\sigma_{el} = 2\pi\int_0^{\pi}(|f|^2 + |g|^2)sin\theta d\theta, \tag{11}$$

and the momentum transfer cross section by

$$\sigma_M = 2\pi\int_0^{\pi}(1 - cos\theta)[|f|^2 + |g|^2]sin\theta d\theta. \tag{12}$$

The integrated total cross section given by

$$\sigma_{tot} = \frac{2\pi}{K^2}\sum_{l=0}^{\infty}\{(l+1)[1-Re(S_l^+)] + l[1-Re(S_l^-)], \tag{13}$$

where $S_l^{\pm} = exp(2i\delta_l^{\pm})$. The integrated absorption cross section can be obtained from $\sigma_{abs} = \sigma_{tot} - \sigma_{el}$

Since the spin-orbit interaction is a short-range interaction, the phase shifts of the spin-up and the spin-down particles are equal ($\delta_l^+ = \delta_l^-$) for the large angular momenta $l\hbar$. Hence for large l, $g(\theta) = 0$ and the contribution to the scattering amplitude comes only from $f(\theta)$. If Born approximation is used for higher partial wave with $l > M$, $f(\theta)$ can be written as [9],

$$f(K,\theta) = \frac{1}{2iK}\sum_{i=0}^{M}[(l+1)(S_l^+ - 1) + l(S_l^- - 1)]P_l + f_B(K,\theta) - \frac{1}{2iK}\sum_{i=0}^{M}(2l+1)(S_{Bl} - 1)P_l, \tag{14}$$

where $f_B(K,\theta)$ is the Born amplitude, $S_{Bl} = exp(2i\delta_{Bl})$ and δ_{Bl} is the Born phase shift. The number of exact phase shifts to be evaluated depends on the impact energy before use of Born approximation. The contribution due to Born approximation should be small. At large distance the interaction potential $V(r)$ is dominated by the long range part $V_{LR}(r) = -\alpha_d/2r^4$ of the polarization potential and Born phase shift δ_{Bl}.

The interaction potential between the spin of the electron or positron and the orbital angular momentum **L**, which depends on the velocity and position vector with respect to the target atom, can cause the spin to orient. Hence, even with an unpolarized incident beam the orientation in a preferred direction can give a net spin polarization in the scattered beam. The amount of polarization produced due to the collision in the scattered beam is given by [182],

$$\mathbf{P}(\theta) = \frac{< A\chi_v|\sigma|A\chi_v >}{< A\chi_v|A\chi_v >} = \frac{f^*g + fg^*}{|f|^2 + |g|^2}\hat{\mathbf{n}} = \mathrm{P}(\theta)\hat{\mathbf{n}}. \tag{15}$$

The other two spin polarization parameters, T and U giving the angle of the component of the polarization vector in the scattering plane are given by [182],

$$T(\theta) = \frac{|f|^2 - |g|^2}{|f|^2 + |g|^2}, \quad U(\theta) = \frac{fg^* - gf^*}{|f|^2 + |g|^2}. \tag{16}$$

The three polarization parameters are interrelated through the condition $P^2 + T^2 + U^2 = 1$.

3. $e^{\pm}$ and Target Atom Interaction Potential

To calculate the scattering parameters, we define the projectile-target interaction and a method to determine the respective wave functions. The scattering can be described in two general categories, elastic (where the total kinetic energy is conserved and the interaction potential is real) and inelastic (where part of the energy is lost due to absorption). For the inelastic processes such as excitation, ionization, positronium formation through electron capture and so forth the absorption potential is developed, which forms the imaginary part of the total complex potential. The total interaction potential between a neutral target (or a single atomic electron) and a projectile electron or positron is assumed to be symmetric or central, $V(r)$, which depends on r only. The general form of $V(r)$ is,

$$V(r) = V_R(r) + iV_A(r), \tag{17}$$

where the real part $V_R(r)$ represents the elastic scattering and the imaginary part $V_A(r)$ represents the absorption of energy through the inelastic channels. When the total kinetic energy is conserved the imaginary part, $V_A(r)$, is zero. The absorption potential is negative and typically depends on the local density function.

$V_R(r)$ has several components: the averaged static potential V_S (attractive for positrons and repulsive for electrons), V_P polarization potential (attractive for both electrons and positrons) and an electron-electron exchange potential V_{ex} (only for electrons). For a positron there is no exchange probability. The total real potential is represented as,

$$V_R(r) = \begin{cases} V_S(r) + V_{P^-}(r) + V_{ex}(r) & : \text{electron scattering,} \\ V_S(r) + V_{P^+}(r) & : \text{positron scattering.} \end{cases} \tag{18}$$

The static potential, V_S, is obtained by averaging the projectile-target interaction over the target wave function as,

$$V_S = \int |\psi_T(\mathbf{r}_1, ..., \mathbf{r}_Z)|^2 \left[\frac{Zee_p}{r} - \sum_{i=1}^{Z} \frac{ee_p}{\mathbf{r} - \mathbf{r}_i} \right] d\mathbf{r}_1 ... d\mathbf{r}_Z = \frac{Zee_p}{r} - \sum_{n,l,m} N_{nlm} \int |\Phi_{nlm}(\mathbf{r})|^2 \frac{ee_p}{\mathbf{r} - \mathbf{r}'} d\mathbf{r}', \tag{19}$$

where ψ_T is the asymmetric Hartree-Fock target wave function, $\Phi_{nlm}(\mathbf{r}) = \phi_{nl}(r) Y_{lm}(\hat{\mathbf{r}})$ are the partial atomic orbitals, e_p is the projectile charge and N_{nlm} is the occupancy number of the orbital (n, l, m). The radial part $\phi(r)$ of an orbital can be an analytic expansion, for example, tables of Clementi and Roetti [183] or in numerical form obtained from configuration interaction atomic structure calculation (e.g., Reference [65] or a Hylleras type wave function expansion (e.g., Reference [26]). Use of configuration interaction form is common in close coupling approximation.

The polarization potential usually has a short and a long range part,

$$V_P(r) = \begin{cases} V_{SR^\pm}(r) \text{ for } r < r_c, \\ V_{LR}(r) = \alpha_o / r^4 \ \text{ for } r \geq r_c, \end{cases} \tag{20}$$

where r_c is the point where the two forms cross each other for the first time. The long range behavior is known to be of the form α_o / r^4 where α_o is the polarizability of the target. The short range form can vary. For the electrons scattering from a neutral atom, it could be the parameter free potential, for example, that given by O'Connel and Lane [184]. They developed the potential on the basis of energy dependent free-electron gas exchange potential and the energy-independent electron-gas correlation potential smoothly joining to the long-range polarization interaction and is given by,

$$V_{SR^-}(r) = \begin{cases} 0.0622 ln r_s - 0.096 + 0.018 r_s ln r_s - 0.02 r_s, r_s \leq 0.7, \\ -0.1231 + 0.03796 ln r_s, 0.7 \leq r_s \leq 10, \\ -0.876 r_s^{-1} + 2.65 r_s^{-3/2} - 2.8 r_s^{-2} - 0.8 r_s^{-5/2}, 10 \leq r_s, \end{cases} \tag{21}$$

$r_s = [3/(4\pi\rho(r))]^{1/3}$, $\rho(r)$ is the undistorted electronic charge density of the target. $\rho(r)$ for the spherically symmetric atom is given by,

$$\rho(r) = \frac{1}{4\pi} \sum_n \sum_l N_{nl} |\phi_{nl}(r)|^2, \tag{22}$$

where N_{nl} is the occupancy number of the orbital (nl).

For the positrons scattering the polarization potential can also be parameter free, such as that by Jain [185]. It is based on correlation energy of a single positron in a homogeneous electron gas with an asymptotic behavior of the long range polarization potential, and is given by,

$$V_{SR^+}(r) = \begin{cases} [-1.82/\sqrt{r_s} + (0.051 ln r_s - 0.115) ln r_s + 1.167]/2, r_s \leq 0.302, \\ (-0.92305 - 0.09098/r_s^2)/2, 0.302 \leq r_s \leq 0.56, \\ [-\frac{8.7674 r_s}{(r_s+2.5)^3} + \frac{(-13.151+0.9552 r_s)}{(r_s+2.5)^2} + \frac{2.8655}{r_s+2.5} - 0.6298]/2, 0.56 \leq r_s \leq 8.0. \end{cases} \tag{23}$$

The long range form of $V_{p^\pm}(r)$ is given by $V_{LR}(r) = -\alpha_d/(2r^4)$, where α_d is the static electric dipole polarizability. In polarized orbital method, the distortion due to polarization is incorporated in the wave function.

The exchange potential, $V_{ex}(r)$ is due to exchange between the projectile electron and the target electrons. One of the common form is given by Riley and Truhlar [186],

$$V_{ex}(r) = \frac{1}{2}\left\{[E - V_D(r)] - \sqrt{(E - V_D)^2 + \rho(r)^2}\right\}, \tag{24}$$

where $V_D = V_S + V_{P^-}$ is the direct interaction potential and $\rho(r)$ is the radial density of the target. Chen et al. [187] introduced another type of potential which was used for elastic scattering from heavy inert gas, Kr, with reasonable success.

When the impact energy becomes accessible for the inelastic processes (such as excitations of the target, positronium formation, etc.) absorption potential is introduced. The total absorption of energy has been represented by various model potentials with poor to good success for certain atoms (e.g., for Ar [9]). One major issue was the inclusion of various threshold energies for excitations and electron capture in case of positrons to form positronium. One successful absorption potential model, especially for electron scattering, has been the semi-empirical potential of Staszewska et al. [188,189], which are based on qualitative features of an absorption potential at short and long ranges in order to predict accurate differential cross sections. Their later model [189] has been in use considerably, and is given by,

$$V_A = -\frac{1}{2}v\rho(\mathbf{r})\bar{\sigma_b}, \quad v = \sqrt{\frac{2(E - V_R)}{m_o}}, \tag{25}$$

v is the local velocity of the projectile for $(E - V_R) \geq 0$, ρ is the target electron density per unit volume and $\bar{\sigma_b}$ is the average quasifree binary cross section for Pauli allowed electron-electron collisions and is obtained non-empirically by using the free-electron gas model for the target as,

$$\bar{\sigma_b}(r, E) = \begin{cases} \frac{32\pi^2}{15p^2}\frac{3}{4\pi k_F^3}\left[\frac{5k_F^3}{\alpha - k_F^2} - \frac{k_F^3[5(p^2-\beta)+2k_F^3]}{(p^2-\beta)^2} + f_2\right], & p^2 \geq \alpha + \beta - k_F^2, \\ 0, \ p^2 < \alpha + \beta - k_F^2, \end{cases} \tag{26}$$

where

$$p(E) = (2E)^{1/2}, \quad f_2(r, E) = \begin{bmatrix} 0, \ p^2 > \alpha + \beta \\ \frac{2(\alpha+\beta-p^2)^{5/2}}{(p^2-\beta)^2}, \ p^2 \leq \alpha + \beta \end{bmatrix}, \quad k_F = (3\pi^2\rho)^{1/3} \tag{27}$$

p is the incident momentum of the projectile and k_F is the target Fermi momentum. In their third version of V_A, V.3, they define the parameters α and β as,

$$\alpha = k_F^2 + 2[\Delta - (I - \Delta)] - V_R, \ \beta = k_F^2 + 2(I - \Delta) - V_R, \tag{28}$$

where Δ is the threshold energy for inelastic scattering and I is the ionization potential. The factor $1/2$ in the equation is introduced to account for the exchange between the incident electrons and the atomic electrons of the target. The same absorption potential can be used for the positron scattering with the factor $1/2$ removed, since there is no exchange effect during the positron scattering. The earlier version of Staszewska et al. [188] has also shown fair to good representation of absorption potential in reproducing the total cross sections. Various other absorption potential models are also available in literature, but have been only partly successful and hence need improvement.

Figure 1 demonstrate the general features of various components of the real part $V_R(r)$ of the total projectile-target interaction potential $V(r)$. The components are static potential (repulsive for electron and attractive for positrons), polarization potential (attractive for both $e^\pm$), exchange potential (only for electrons and attractive) for $e^\pm$ scattered by the cadmium atoms [8]. The static potential was obtained using Slater-type orbitals of Roothan-Hartree-Fock wave functions given by Clementi and Roetti [183].

The same orbital functions were used to obtain the electron density in the absorption potentials. As expected, the static potential dominates near to the nucleus and exchange potential starts away from it, but moves toward it with increasing energy of the projectile.

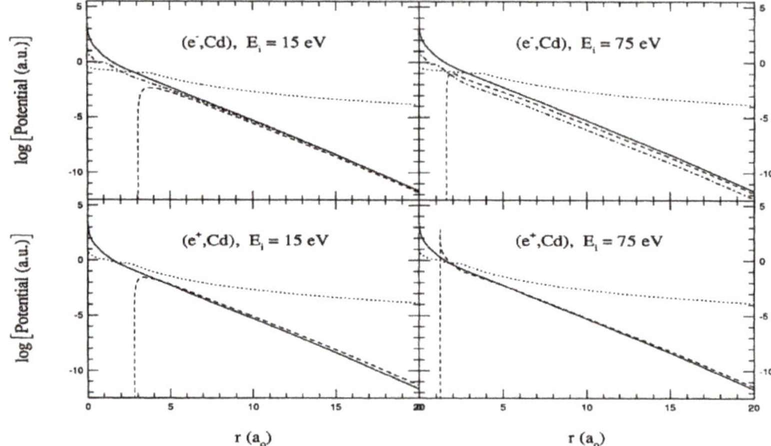

Figure 1. Various interacting potentials, V_S (solid), V_P (dotted), V_{ex} (dashed) for $e^{\pm}$ scattering from Cd showing behaviors at energies of 15 and 75 eV. The curves represent the absolute values of the potentials which are all negative except V_S which is positive for electrons [8].

4. $e^{\pm}$ Scattering from Molecules

Two of the most commonly used approaches to study electron/positron interaction from molecules are described below:

4.1. Optical Potential Approach

As mentioned above, the positron-target interaction is represented by an optical potential, which has a real part representing elastic processes and the imaginary part taking care of the loss of flux due to different inelastic channels. The calculations are generally carried out under the fixed nuclei (FN) approximation. This method has been successful in investigating not only atoms, but also larger molecules under multicentre approach. Various additivity rules are applied to find the cross section of a molecule from its atomic constituents. The independent atom model (IAM) is the simplest among them, where the cross section of the constituent atoms of the molecule is simply added to find the cross section for the molecule [72]. This approach was further modified and screening correction (IAM-SCAR) was applied to the cross section including the interference term and rotational excitation cross sections [190].

The methodology to obtain radial charge density and interaction potentials for atomic systems was already described earlier. The parametric form for atomic ρ and V_{st} are then used to compute electron/positron impact cross section for atoms in the spherical complex optical potential (SCOP) approach [49]. In this method, the radial part of Schrödinger equation is solved through partial wave analysis to obtain the asymptotic solution,

$$u_l(k,r) \xrightarrow[r\to\infty]{} A_l(k)\sin\left[kr - \frac{1}{2}l\pi + \delta_l(k)\right], \tag{29}$$

where $\delta_l(k)$ is the phase shift, which carries the signature of interaction on the solution. To compute phase shift numerically, the radial Schrödinger equation is solved in the region $r < a$, where a is the finite range of the interaction potential. By applying the boundary condition, $r = a$ we have,

$$R_l(k,r) = A_l(k,r)[j_l(k,r) - \eta_l(k,r)\tan\delta_l]_{r=a}, \tag{30}$$

and its logarithmic derivative,

$$\gamma_l = \left[R_l^{-1}\left(\frac{dR_l}{dr}\right)\right]_{r=a} = k\left[\frac{j_l'(k,r) - \eta_l'(k,r)\tan\delta_l(k)}{j_l(k,r) - \eta_l(k,r)\tan\delta_l(k)}\right]_{r=a}, \tag{31}$$

where j_l and η_l are the spherical Bessel and Neumann functions respectively. The 'prime' means derivative with respect to 'r'. By inverting the equation as above we can obtain,

$$\tan\delta_l(k) = \frac{kj_l'(k,a) - \gamma_l(k)j_l(k,a)}{k\eta_l'(k,a) - \gamma_l(k)\eta_l(k,a)}. \tag{32}$$

The phase shift, $\delta_l(k)$ obtained from above is then used to construct elastic and inelastic cross sections as [182],

$$Q_{el}(k) = \frac{\pi}{k^2}\sum_{l=0}^{\infty}(2l+1)|\eta_l \exp(2iRe(\delta_l)) - 1|^2, \quad Q_{inel}(k) = \frac{\pi}{k^2}\sum_{l=0}^{\infty}(2l+1)\left(1-\eta_l\right)^2. \tag{33}$$

However, for molecules there is no direct method to obtain ρ and V_{st} to be used to compute cross sections. Hence, various indirect techniques were adopted to approximate the charge density and static potential of molecules. For closed molecules like CH_4, a single center expansion (SCE) is used, where the charge density of lighter atom is expanded from the center of mass of the heavier atom [191,192]. However, for larger non-spherical molecules, this is not possible. Antony and co-workers [50,193,194] have extensively used group additive rule to efficiently compute molecular charge density (and static potential). In this method the molecule is assumed to be made of different centers/groups having a group of atoms, each of them scattering positron independently. The grouping of atoms are performed based on the assumption that the charge density of one group do not overlap with the other. Even though such an approximation is a simplification of the actual scenario, it is sufficient to produce reasonable results in the present case (intermediate to high energy projectiles) without the loss of generality. Thus, for each scattering center the ρ and V_{st} are developed and the desired cross section is computed for that center. Finally, cross sections calculated from each scattering center is added to get the molecular cross section. The total ionization cross section Q_{ion} is obtained from the Q_{inel} using complex scattering potential-ionization contribution (CSP-ic) technique [195,196]. Further details on the method is presented in the references.

In 1996 Reid and Wadehra [197] introduced an absorption model for positron collisions. The form of this potential is given as,

$$V_{abs} = -\frac{1}{2}\rho(r)v_{loc}\sigma_{pe}, \quad \sigma_{pe} = 4\pi\left(\frac{a_0 R}{\epsilon E_F}\right)^2 \begin{cases} f(1); & \epsilon^2 - \delta \geq 1 \\ f\left(\sqrt{\epsilon^2 - \delta}\right); & 1 \geq \epsilon^2 - \delta \geq 0 \\ f(0); & 0 \geq \epsilon^2 - \delta \end{cases} \tag{34}$$

In the above equation, a_0 is the Bohr radius and R is the Rydberg constant, $\delta = \frac{\Delta}{E_F}$, $\epsilon = \sqrt{\frac{E_i}{E_F}}$, $f(x) = \frac{2}{\delta}x^3 + 6x + 3\epsilon\ln\frac{\epsilon - x}{\epsilon + x}$ and E_F is the Fermi energy. Δ defines a threshold below which all inelastic processes such as excitation and ionization are energetically forbidden. Determination of Δ is important while dealing with positrons. Reid and Wadehra [197] in their original model used $\Delta = \Delta_p = I - 6.8$ eV (where Δ_p is the Ps formation threshold energy). However, this cannot be used as a general rule, since excitations may occur below this energy. The parameter free model developed to find Δ is given as [198],

$$\Delta = \Delta_e - \frac{(\Delta_e - \Delta_p)}{1 + \left(\frac{E_i}{5I} - \frac{1}{3}\right)^3}, \tag{35}$$

where E_i is the incident energy and Δ_e is the lowest electronic excitation energy.

4.2. *ab initio* R-Matrix Method

The R-matrix method was developed to study scattering problems in nuclear physics by Wigner and others [199,200] in the late 1940s. Burke et al. [201] modified it to include electron-atom scattering processes and subsequently adapted to carry out accurate calculations for diatomic molecules [202]. This *ab initio* theory is primarily based on the concept of constructing two divisions in coordinate space (inner and outer regions) to adequately and efficiently represent the target and scattering regions. These regions are separated by a spherical boundary of radius 'a' containing the target molecule with N + 1 indistinguishable electrons (where +1 is due to the scattering electron). This N + 1 collision system acts as a bound state, where the short-range correlation effect and electron exchange among the electrons are dominant. This makes the inner region scattering problem quite complex. However, calculation is needed to be done only once as the problem is independent of energy. For an accurate solution of the inner region, Quantum chemistry codes are employed generating an energy independent wave function for the N + 1 electrons system. The results obtained in the inner region are fed to the outer region, which contains only the scattering electron. In the outer region, the scattering electron is far away from the center and thus exchange and correlation between electrons can be ignored. Therefore, only the long range multipolar interaction between the scattering electron and the target electron is considered here. For the simplicity of the numerical problem, single center approximation is employed, which converges quite rapidly without losing the generality of the problem. Since the calculations in the outer region are relatively simple, they are repeated for each electron energy on a finite grid to find desired scattering properties such as cross sections. In general, the inner region radius is varied from around 10 to 15 a.u. depending on the size of the target, while the outer region is kept infinite (100 a.u.). The value of R-matrix radius is chosen to accommodate total wave within the sphere.

The molecular orbitals are developed from atomic ones, which are expressed as basis functions centered on respective nuclei. The orbitals are a representation of the molecular charge density distribution, which must be negligibly small at the R-matrix boundary. The electrons are assumed to be in certain combinations of target orbitals to produce configuration state functions (CSF) in various total symmetries. A configuration interaction (CI) molecular wave function is expressed as a linear combination of CSFs in CI expansion. The molecular orbital representation in occupied and virtual orbitals are constructed using Hartree-Fock Self-Consistent Field (HF-SCF) method. For this, Gaussian-type orbitals (GTO) are used and the continuum orbitals of Faure et al. [203] are included up to g ($l = 4$). For dipole-forbidden excitations, $\Delta J \neq 1$, J is rotational constant without spin coupling, the convergence is rapid for partial waves. However, for dipole-allowed excitations, $\Delta J = \pm 1$, convergence is quite slow due to the long range dipole interactions.

The closed coupling expansion (CC) of inner region wave function under fixed nuclei approximation is expressed,

$$\Psi_k = A \sum_{ij} \Phi_i(x_1, ..., x_N) u_{ij}(x_{N+1}) a_{ijk} + \sum_i \chi_i(x_1, ..., x_{N+1}) b_{ik}, \tag{36}$$

where 'A' is the anti-symmetrization operator. $u_{ij}(x)$ are target continuum orbitals and $\Phi_i(x)$ are target wave functions (together called 'target+continuum' configuration). Φ_i are constructed from basic HF method and CI expansion is used for the other. The χ_i are two-center quadratically integrable (L^2) functions. These are formulated with target occupied and virtual orbitals. Care has to be taken in choosing the correct basis function as final results are strongly depended on it. Since the continuum orbitals do not vanish at the R-matrix boundary ($r = a$), the Hamiltonian is modified by adding a surface or Bloch operator [204] to make it Hermitian inside the R-matrix sphere. The form of Bloch operator is,

$$L_{N+1} = \frac{1}{2}\sum_{i=1}^{N+1}\sum_{j}\left[\psi_j^N Y_{l_j m_j}(\hat{r}_i)\delta(r_i - a)\left(\frac{dr}{dr_i} - \frac{b-1}{r_i}\right)\right]\psi_j^N Y_{l_j m_j}(\hat{r}_i). \tag{37}$$

Then by diagonalizing the modified Hamiltonian $(H_{N+1} + L_{N+1})$ the inner region electron wave function is obtained. The inner region solutions are used to set up the R-matrix at the boundary of the sphere. Then it is propagated to far region (a_{out}), where it is matched with the asymptotic functions obtained from the Gailitis expansion [205]. The numerical value of a_{out} depends on the nature of the target. The quality of any scattering model depends on a satisfactory representation of the target. Therefore getting the right values of target parameters viz. ground state energy, vertical excitation energies, ground state dipole moment and rotational constant is a necessary precursor to scattering calculations.

The exchange and correlation effects are negligible in the outer region and the physical interactions dominate the scattering processes. The scattering electron moves under the influence of long range multipole potentials of the target. Hence, a single center expansion of the scattering wave function sufficient here. This is given by,

$$\psi = \sum_i \bar{\phi}_i(x_i ..., \sigma_{N+1}) r_{N+1}^{-1} F_i(\hat{r}_{N+1}) Y_{l_j m_j}(\hat{r}_{N+1}), \tag{38}$$

where $x_j = (\hat{r}_j, \sigma_j)$ is the position and spin of the jth target electron, the functions $\bar{\phi}_i$ are formed by coupling the scattering electron spin σ_{N+1} with the target state ϕ_i and F_i are reduced radial wave functions. By appropriate substitutions, a set of coupled, homogeneous, differential equations for the reduced radial wave function are obtained. The solution of this equation is obtained by propagating the R-matrix from boundary to sufficiently large distances, where the interaction between scattering electron and target molecule tends to zero [206]. Thereafter, asymptotic expansion techniques are used to solve for the outer region functions [207]. In the limit $r \to \infty$ above equation have different linearly independent standing wave asymptotic solution j for each energetically open channels i given as,

$$F_{ij} \cong \frac{1}{\sqrt{k_i}}\left[\sin\left(k_i r - \frac{1}{2}l_i\pi\right)\delta_{ij} + \cos\left(k_i r - \frac{1}{2}l_i\pi\right)\right]K_{ij}. \tag{39}$$

The coefficients K_{ij} define the real, symmetric K-matrix, which contains all the scattering information. The eigenphase sum δ is used for the detection and parametrization of resonances, obtained directly from the diagonalized K-matrix, K_{ij}^D as $\delta = \sum_i \arctan\left(K_{ij}^D\right)$, where the summation is over the open channels. The scattering matrix S is a transformation of the K-matrix given by, $S = (1 + iK)(1 + ik)^{-1}$.

The integral cross section for the excitation from states i to i' [208] is given as,

$$\sigma\left(i \to i'\right) = \frac{\pi}{k_i^2}\sum_S \frac{2S+1}{2(2S+1)}\sum_{\Gamma l l'}|\Gamma_{i l i' l'}^{\Gamma S}|^2, \tag{40}$$

where S_i is the spin angular momentum of the ith target state, S is the total spin angular momentum, Γ runs over the symmetry and l and l' are orbital angular momentum quantum numbers corresponding to i to i' states respectively. K-matrices form the input of POLYDCS program [209], from which the scattering observables are evaluated. For further details on R-matrix methodology to study positron collisions, please refer to References [115,118] and references therein.

5. Results and Discussions

Positron and electron scattering from neutral atoms/molecules and one or two electron ions have been investigated extensively as seen in the partial list of references in this article. For consistent features and feasible experimental set-ups, the targets have been mainly hydrogen atom, inert gases, He^+ and $Li^{+,2+}$ ions, symmetric closed shell molecules, and few other simple molecules. Theoretical methods have been developed with significant accuracy for benchmarking the experiments. However, issues remain unresolved for better representation of the interaction and the wave function for sensitive cases. Examples of various features of the scattering cross sections and spin polarization for $e^{\pm}$ from atoms and molecules are illustrated below.

5.1. Cross Sections and Spin Polarization for $e^{\pm}$ Scattering from Atoms

Since the theoretical basis for $e^{\pm}$ scattering from atoms is similar, reproduction of electron scattering cross sections gives a measure of accuracy of the method to study positron scattering [210,211]. However, its worth noting that the presence of Ps formation channel strongly affects the shape of the cross section in the energy range from its threshold to around 200 eV. Figure 2 left panel gives an example of accuracy in the features of differential cross sections (DCS) for scattering of electrons from argon, where measured DCS from 7 different experiments [18–20,22–24,212] are compared with theoretical predictions in nonrelativistic [1] and relativistic [9] approximations. Nahar and Wadehra obtained the static potential using analytic orbitals obtained by Clementi and Roetti [183] and absorption potential from Staszewska et al. [189]. Very good agreement in features, shape and magnitude benchmarks the theory. It may be noted that use of absorption potential has little effect on the DCS values as expected for elastic scattering and that relativistic corrections have removed some discrepancy at the minimum for DCS at 300 eV.

The potentials used for the cross sections, were used to study the angular dependence of spin polarization of the elastically scattered electrons from argon as demonstrated on the right panel of Figure 2 [9] at several electron impact energies. The spin-orbit polarization is highly sensitive to interaction potential. We see very good agreement between theory and experiment at 40 eV projectile energy. However, comparison between the theoretical and experiment values show only similar qualitative features at 50 and 100 eV. Spin polarization can be a sensitive accuracy indicator for the benchmark of both theory and experiment.

Figure 3 [9] presents the integrated relativistic cross section, where the left panel shows the elastic and the right panel shows total scattering cross section for electrons from argon. The trend shows cross section rising from very low energy to form a peak around 15 eV and then decaying. The predicted cross sections from relativistic Dirac equation [9] and from non-relativistic Schrödinger equation [1], both using the optical potential, compare very well with the measured values in the low energy region for both elastic and total scattering. The predicted integrated elastic scattering cross sections using V_R do not show any difference between relativistic and non-relativistic approach. They fall almost on the same curve (solid and dot-dashed) and agree well with the measured values from 6 sets of data from different sources—open triangle [22], cross [20], diamond [23], plus [24], open circle [34], asterisk [35]. Cross sections with inclusion of absorption potential [189] (dotted curve) improves the agreement slightly. However, cross sections with inclusion of an older form absorption potential [188] (dashed curve) show better agreement with couple of experiments, for example, open squares [18]. The earlier version of the absorption potential [188] seems to produce total cross sections closer to the experiments than those by version 3 where the measured values are from [25] (open square), [37] (open circle), [38] (cross), [36] (asterisk), [39] (open triangle).

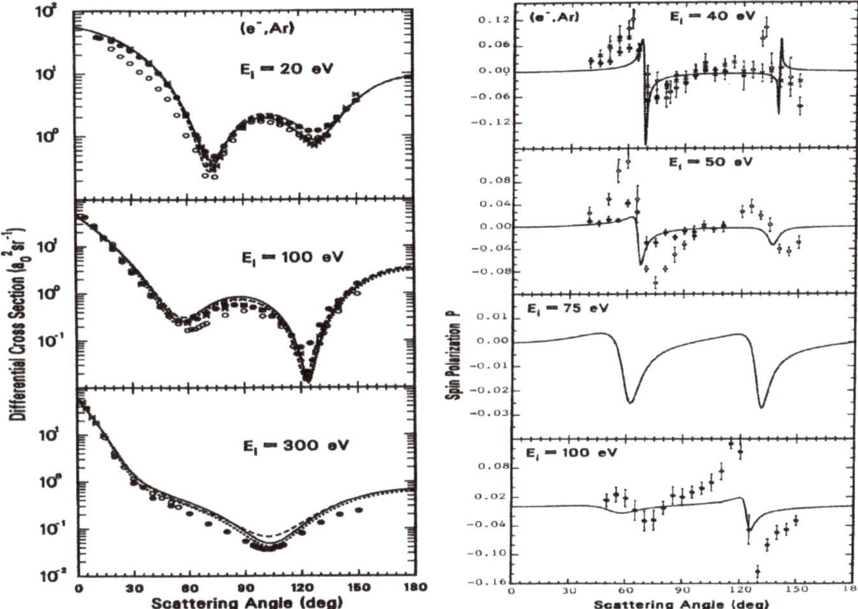

Figure 2. Left: Differential cross sections (DCS) for elastic scattering of electrons from Ar at 20, 100, and 300 eV illustrating features from interference effects and comparing theoretical predictions from optical potential approach with experimental measurements. Theory: solid curve represents use of only real part of the the potential and dot that including both real and absorption potentials [9], dash represents nonrelativistic approach with only real potential [1]). Experimental data corresponds to 7 different measurements: at 20 eV, filled circle [22], open circle [18], at 100 eV, open circle [18], solid circle [20], at 300 eV, filled circle [23], open circle [24], asterisks [19]. **Right**: Angular dependence of spin polarization P for elastic scattering of electrons from Ar at various electron impact energies. Theoretical prediction from optical potential approach [9] is compared with measured values from 3 set-ups—open circles at 40 eV [40], solid circles at 40 and 50 eV [41], cross at 40 eV [42].

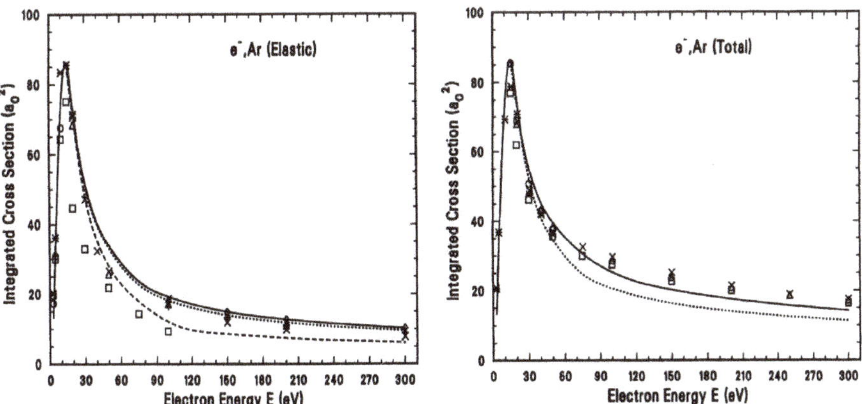

Figure 3. Integrated cross sections for elastic scattering (**left** panel) and total scattering (**right** panel) of electrons from Ar at various energies. A rising broad peak in cross section in the low energy below 30 eV is visible [9]. Comparison show good agreement as explained in the text.

Number of measurements for elastic and total scattering of positrons from atoms are relatively lower than that of electrons. Differential cross sections for positron scattering from argon was first studied experimentally by Hyder et al. [27]. The theoretical study for DCS by Nahar and Wadehra [9] using the the optical potential approach described in the theory section was compared (Figure 4) with this single measurement available at that time of reporting. To compare the impact of various absorption potentials, the same set of experimental values was normalized at 90° with different symbols to differentiate the type of curves produced by different absorption potentials. The comparison shows that the relative features of the measured DCS values agree very well with the predicted values when only the real potential (solid curve) or the complex potential that included version 3 of absorption potential given by Staszewska et al. [189] (dotted curve) were used. It may be noted that in contrast to electron scattering, positron scattering shows less interference structures and decays smoothly with impact energy. Use of other absorption potentials introduce dips and humps in the DCS curves not seen in the experiment.

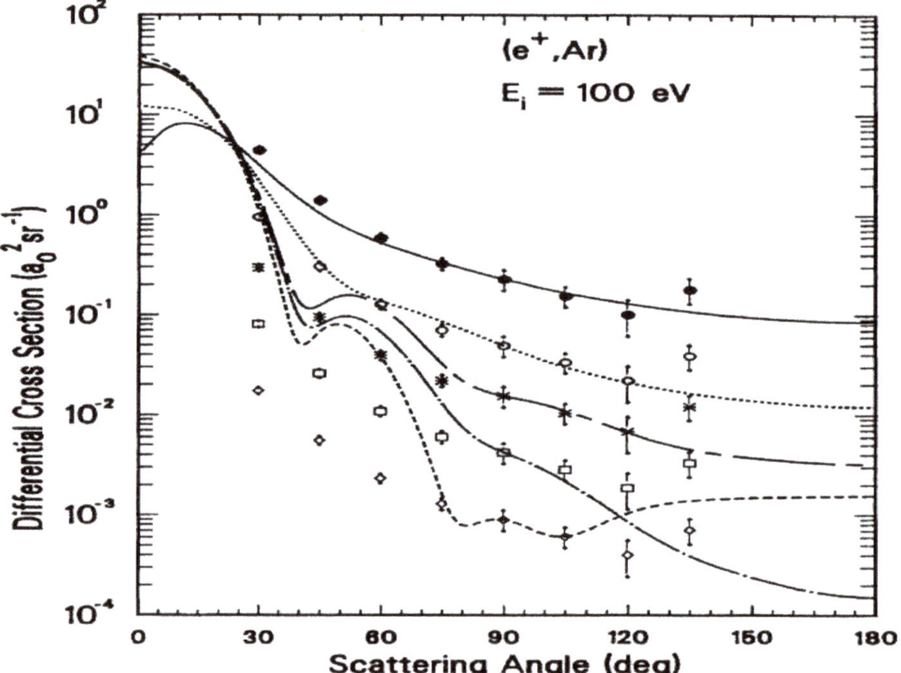

Figure 4. Differential cross sections (DCS) for elastic scattering of positrons from Ar at 100 eV illustrating features and comparison between theory [9] and experiment [27]. The experimental values have been normalized at 90° with different symbols to differentiate curves to compare the differences in DCS values with different absorption potentials.

Figure 5 left panel presents total integrated cross sections for positrons scattering from Ar atoms where theoretical cross sections are compared with measured values from 4 experiments. Although the measured relative values of DCS agreed well with predicted features obtained using the version 3 absorption potential of Reference [189], measured integrated cross sections are seen to be higher than the predicted values (dotted curve). Use of their earlier version of the absorption potential [188] yielded cross sections (solid curve) that show better agreement in the lower energy, but remains higher overall at higher energies. These indicate that absorption potentials for positron scattering have partial success in representing the positron-atom interaction. The right panel presents a later study of Jones et al. [26] who carried out both the experiment and calculated the total cross sections for positrons scattering from argon and neon. They claim significant improvement in agreement between the two. They measured the grant total cross sections (σ_{GT}) which includes excitations and formation of positronium from zero to 60 eV in contrast to earlier existing experiments, where the energy goes up to 300 eV. They do not show the broad rising peak at low energy as seen Figure 5. Their numbers for the cross sections appear to be different from those in the left panel. The reason for this difference is not clear. For the theoretical predictions, they consider two approaches (i) relativistic optical potential (ROP) method, which reduces to their relativistic polarized orbital method below the first excitation threshold and (ii) convergent close coupling (CCC) method. Their ab initio absorption potential correspond to inelastic scattering due to excitations and positronium formation channels. There is an overall improvement of their approaches in agreeing with the measured values. It is difficult to determine which approach worked better, since their CCC shows better agreement at lower energies with their own experiment, while ROP is better with other measurements.

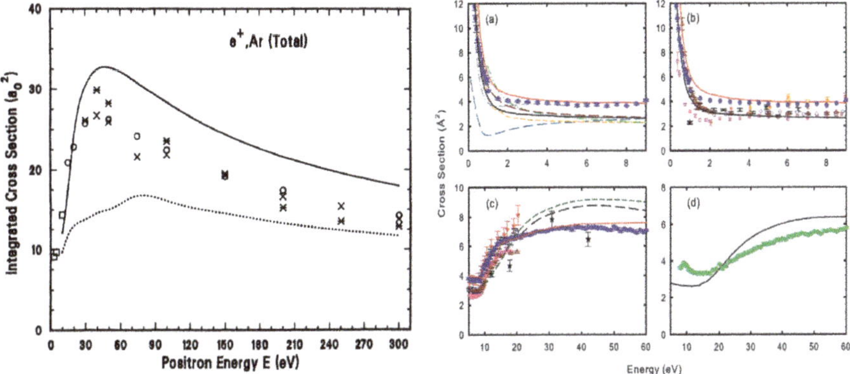

Figure 5. **Left**: Comparison of total integrated cross sections for scattering of positrons from Ar at various energies between the theory using optical potential [9] and 4 experiments—open squares [21], open circles [25], asterisks [28], and cross [29]. The solid curve corresponds to use of absorption potential of Reference [188] and dotted curve to version 3 potential of Reference [189]. Although sold curve shows agreement in the low energy cross sections the rest of the measured integrated cross sections remain in-between the two predictions. **Right**: Total integrated cross sections for scattering of positrons from Ar at various energies by Jones et al. [26]: (**a**) Measured grand total cross section (σ_{GT}) (filled circles) below the ionization threshold for positronium formation (E_{Ps} are compared with various theoretical approximations, (**b**) Measured σ_{GT} is compared with other experiments, (**c**) Measured σ_{GT} above (E_{Ps} are compared with other measured values and theories, (**d**) Measured cross section σ_{GT-Ps} where cross section for positronium has been subtracted above (E_{Ps} are compared with their ROP prediction.)

In contrast to Ar, optical potential approach has shown better agreement with experiments in reproducing the integrated elastic scattering cross sections for Mg (e.g., Reference [10]) and total scattering cross sections for Na atoms (e.g., Reference [12]) by positrons. The left panel of Figure 6 shows integrated cross sections for elastic scattering of positrons from magnesium by Hossain et al. [10] (solid curve) using an optical potential similar to that used for Ar by Nahar and Wadehra [9] and obtained very good agreement with the measured values of Stein et al. [32,33]). The right panel in the figure presents total cross sections for positron scattering from Na atom [12] where optical potential approach [12] was found to be higher in magnitude than other results, but with similar shape and within the experimental errors of Kaupilla et al. [30] and Kwan et al. [31]. The cross-sections of Reid and Wadehra [13], Hewitt et al. [14] and Lugovskoy et al. [15] falls below the SCOP data. However, the data of Cheng et al. [16] show a different nature, even though falls very close to other values.

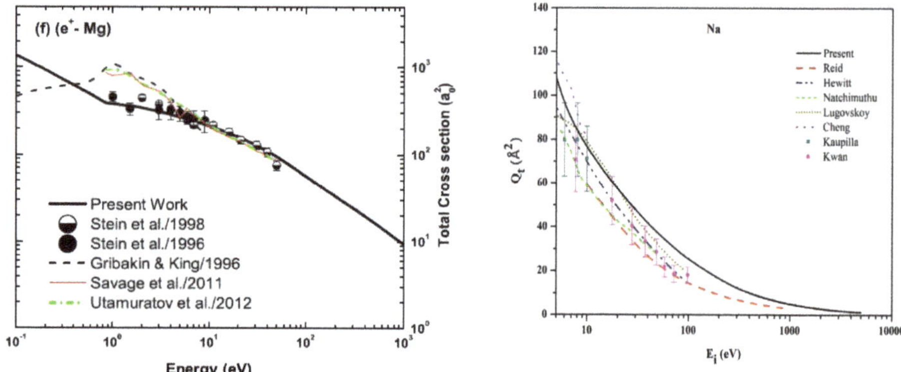

Figure 6. Left: Comparison of predicted total integrated cross sections for scattering of positrons from magnesium obtained by Hossain et al. [10] using optical potential similar to the present work show good agreement with experimental values. **Right**: Total cross section for e^+-Na interaction [12].

5.2. $e^{\pm}$ *Scattering from Molecules*

We present some illustrative examples of electron and positron scatterings from molecules. Figure 7 presents positron impact total cross section from HCl molecule. The result obtained employing the SCOP method is compared with the theoretical values of Sun et al. [213] and measurements of Hamada et al. [214]. The theories agrees reasonably well in the high energy range. However, below 40 eV the data of Sun et al. [213] underestimate SCOP. This is probably because they performed the calculation using additivity rule to compute the total cross sections for the molecule. Further, they have taken Δ_p as the inelastic threshold, which results in larger magnitude of cross sections at low energies. On the other hand, SCOP values show excellent agreement the measurements of Hamada et al. [214], both in shape and magnitude. Their values are corrected for forward angle scattering, which strengthen our case in terms of reliability of data obtained using SCOP method.

Figure 8 shows Q_{tot} for electron/positron scattering by $C_4H_4N_2$ molecule. For both cases SCOP method is in very good agreement with the theoretical values reported by other groups [93,94,215,216]. However, for positron scattering data of Sanz et al. [93] falls faster that other cross sections. The measurements [217–219] also show reasonable agreements with theories, except with Reference [220] for electron scattering and Reference [94] for positron scattering. Both these measurements underestimates other values significantly [221].

Figure 9 shows the comparison of electron/positron scattering total and ionization cross sections respectively for propene. In Figure 10 the ratios of cross section is plotted to clearly show the variation in cross section. As expected, the total cross section for electrons is larger than positrons in the intermediate energies. This is very clear in the ratio in Figure 10. However, for the ionization curve, the trend is reversed at intermediate energies. At high energies, typically around 1 keV, both curves tends to converge. This is because at such energies the scattering potential is weak compared to the kinetic energy of the projectile and the scattered wave is approximated as plane wave, in accordance with the first Born approximation.

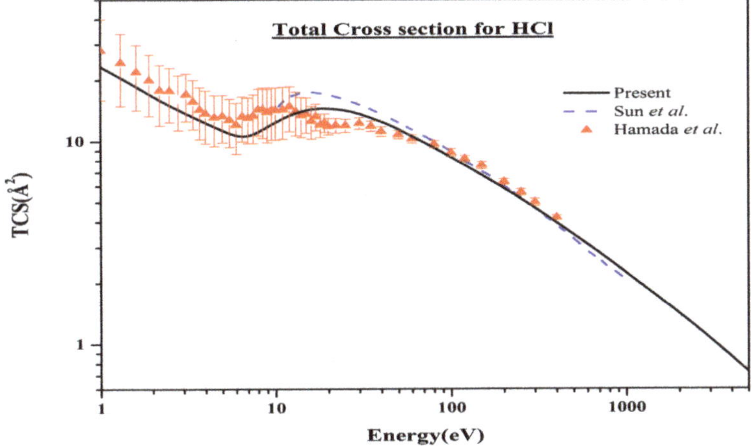

Figure 7. Total cross section for e$^+$-HCl molecule interaction [50].

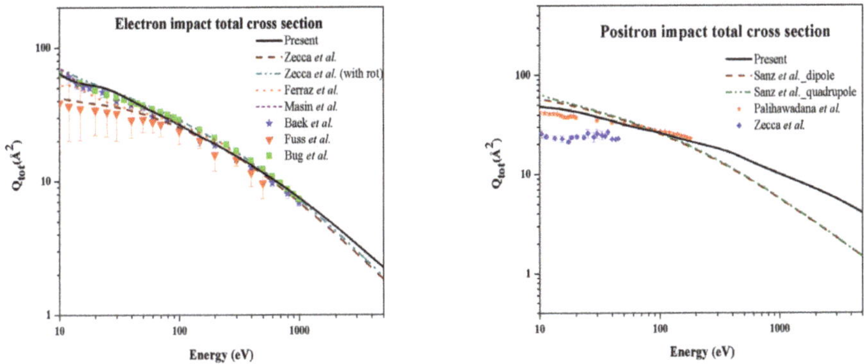

Figure 8. Left: Total cross section for e$^-$-pyrimidine molecule interaction [222]. **Right**: Total cross section for e$^+$-pyrimidine molecule interaction [222].

Figure 9 shows the comparison of electron/positron scattering total and ionization cross sections respectively for propene. In Figure 10 the ratios of cross section is plotted to clearly show the variation in cross section. As expected, the total cross section for electrons is larger than positrons in the intermediate energies. This is very clear in the ratio in Figure 10. However, for the ionization curve, the trend is reversed at intermediate energies. At high energies, typically around 1 keV, both curves tends to converge. This is because at such energies the scattering potential is weak compared to the kinetic energy of the projectile and the scattered wave is approximated as plane wave, in accordance with the first Born approximation.

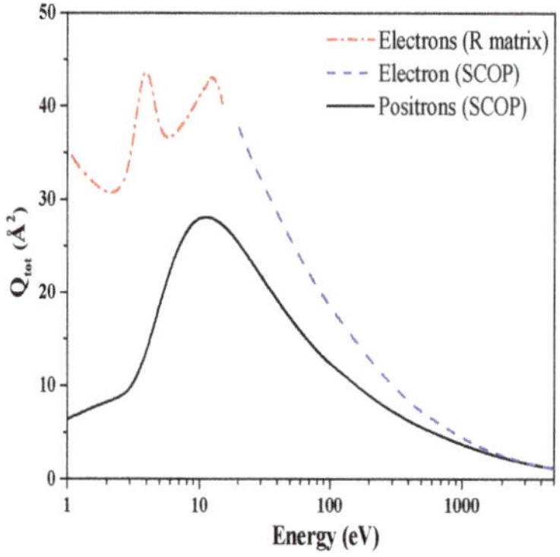

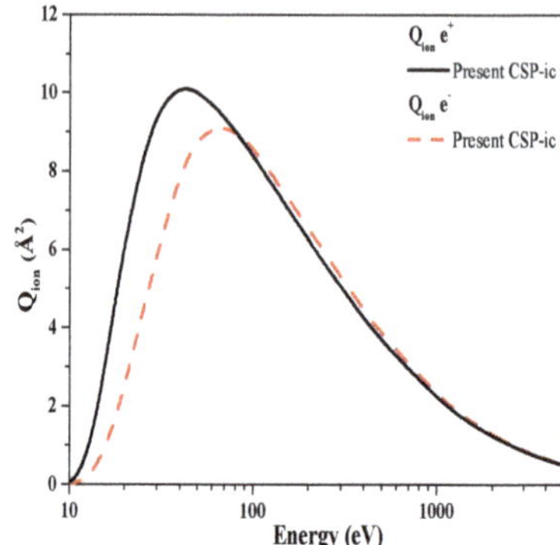

Figure 9. Left: Total cross section for e^-/e^+-propene interaction [223]. **Right**: Total ionization cross section for e^-/e^+-propene interaction [223].

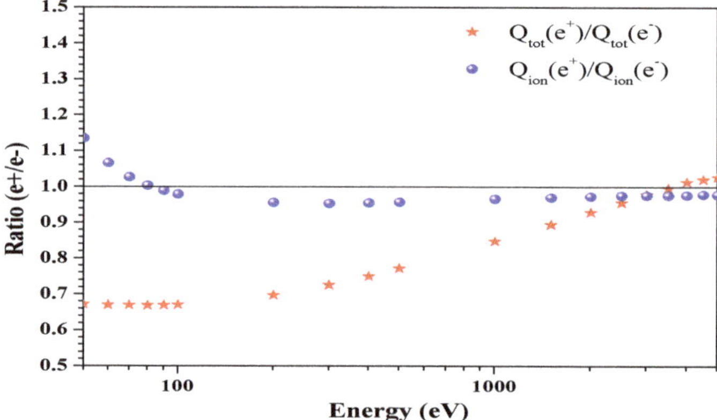

Figure 10. Ratio of cross section for e^-/e^+-propene interaction [223].

Positron impact total cross sections are computed for many atomic and molecular targets by Antony and his group using SCOP formalism. The list of the targets studied, with references to articles published, are presented in the Table 2.

Table 2. Positron scattering cross section predicted through optical potential method with references (Q_{tot}: Total CS, Q_{ps}: positronium formation CS, Q_{el}: elastic CS, Q_{ion}: direct ionization CS, Q_{mtcs}: momentum transfer CS, Q_{inel}: ine;elastic CS, Q_{tion}: total ionization CS, Q_{diff}: differential CS).

Target	Cross Sections	Reference
Inert gases	Q_{tot}, Q_{ion}, Q_{ps}, Q_{tion}	[11]
C, N, O	Q_{tot}, Q_{ion}, Q_{ps}, Q_{tion}	[71,198]
Be, Mg, Ca, Sr, Ba, Ra	Q_{tot}, Q_{ps}, Q_{el}, Q_{ion}, Q_{mtcs}, Q_{inel}, Q_{tion}	[224]
Li, Na, K, Rb, Cs, Fr	Q_{tot}	[12]
C_2, N_2, O_2	Q_{tot}, Q_{ion}, Q_{ps}, Q_{tion}	[71,198]
CH_4, CO, CO_2, H_2, N_2, NO	Q_{tot}, Q_{ion}, Q_{ps}, Q_{tion}	[196,225]
H_2O, NH_3, HCl, OCS, SO_2	Q_{el}, Q_{tot}	[50]
SiH_4, GeH_4, SnH_4, PbH_4	Q_{ion}, Q_{ps}, Q_{tion}	[226]
CH_3F, CH_3Cl, CH_3Br, CH_3I	Q_{tot}, Q_{ion}, Q_{ps}, Q_{tion}	[227]
C_2, C_2, C_2H_2, C_3H_8, C_3H_4	Q_{tot}, Q_{el}, Q_{ion}, Q_{ps}, Q_{tion}	[228]
C_3H_6	Q_{tot}, Q_{ion}	[223]
n-pentane, iso-pentane, neo-pentane	Q_{el}, Q_{inel}, Q_{diff}, Q_{tot}	[229,230]
$C_4H_4N_2$	Q_{el}, Q_{tot}	[222]

6. Conclusions

We present a review of the elastic and inelastic scattering of electrons and positrons from atoms and molecules. The present conclusion can be given by the following points:

- Current theoretical methods for electron scattering from inert atoms are well established. However, their predictions for spin polarization may provide a measure of their effective representation.
- Current theoretical method for elastic scattering of positrons from inert gases, single and double electrons systems and molecules have also been developed well, but further studies are needed.
- There is a critical need for improved absorption potential for positron scattering from atoms and molecules.
- Further study is needed to expand the scope for scattering from ions that are abundant in astrophysical plasmas.

Author Contributions: S.N.N. and B.A. have written up the general background for electron-positron scattering from atoms and molecules. S.N.N. has presented the review of the electron-positron scattering from atoms while B.A. presented those from molecules. All authors have read and agreed to the published version of the manuscript.

Funding: This research received no external funding.

Acknowledgments: S.N.N. acknowledges partial support from the donation in kind grant to Astronomy Department of the Ohio State University, and NSF AST-1312441.

Conflicts of Interest: The authors declare no conflict of interest.

References

1. Nahar, S.N.; Wadehra, J.M. Elastic scattering of positrons and electrons by argon. *Phys. Rev. A* **1987**, *35*, 2051–2064. [CrossRef]
2. Temkin, A. Polarization and Exchange Effects in the Scattering of Electrons from Atoms with Application to Oxygen. *Phys. Rev. A* **1957**, *107*, 1004–1012. [CrossRef]
3. Temkin, A. A Note on the Scattering of Electrons from Atomic Hydrogen. *Phys. Rev. A* **1959**, *116*, 358–363. [CrossRef]
4. McEachran, R.P.; Morgan, D.L.; Ryman, A.G.; Stauffer, A.D. Positron scattering from noble gases. *J. Phys. B* **1977**, *10*, 663–677. [CrossRef]
5. Dasgupta, A.; Bhatia, A.K. Scattering of electrons from argon atoms. *Phys. Rev. A* **1985**, *32*, 3335–3343. [CrossRef]
6. Drachman, R.J. Theory of low energy positrion-helium. *Phys. Rev.* **1966**, *144*, 25. [CrossRef]
7. Campeanu, P.I.; Humberston, J.W. The scattering of p-wave positrons by helium. *J. Phys. B* **1975**, *8*, L244. [CrossRef]

8. Nahar, S.N. Cross sections and spin polarization of $e^{\pm}$ scattering from cadmium. *Phys. Rev. A* **1991**, *43*, 2223–2236. [CrossRef]
9. Nahar, S.N.; Wadehra, J.M. Relativistic approach for $e^{\pm}$ scattering from argon. *Phys. Rev. A* **1991**, *53*, 1275–1289. [CrossRef]
10. Hossain, M.I.; Haque, A.K.F.; Patoary, M.A.R.; Uddin, M.A.; Basak, A.K. Elastic scattering of electrons and positrons by atomic magnesium. *Eur. Phys. J. D* **2016**, *70*, 41. [CrossRef]
11. Singh, S.; Naghma, R.; Kaur, J.; Antony, B. Study of elastic and inelastic cross sections by positron impact on inert gases. *Eur. Phys. J. D* **2018**, *72*, 69. [CrossRef]
12. Sinha, N.; Singh, S.; Antony, B. Positron total scattering cross-sections for alkali atoms. *J. Phys. B Atomic Mol. Phys.* **2018**, *51*, 015204. [CrossRef]
13. Reid, D.D.; Wadehra, J.M. Intermediate- to high-energy positrons scattered by alkali-metal atoms. *Phys. Rev. A* **1998**, *57*, 4. [CrossRef]
14. Hewitt, R.N.; Noble, C.J.; Bransden, B.H. Positron collisions with alkali atoms at low and intermediate energies. *J. Phys. B Atomic Mol. Opt. Phys.* **1993**, *26*, 3661–3677. [CrossRef]
15. Lugovskoy, A.V.; Kadyrov, A.S.; Bray, I.; Stelbovics, A.T. Two-center convergent-close-coupling calculations for positron-sodium collisions. *Phys. Rev. A* **2012**, *85*, 034701. [CrossRef]
16. Cheng, Y.-J.; Zhou, Y.-J.; Jiao, L.-G. Total cross sections of positron sodium scattering at low energies. *Chin. Phys. B* **2012**, *21*, 013405. [CrossRef]
17. Canter, K.F.; Coleman, P.G.; Griffith, T.C.; Heyland, G.R. The measurement of total cross sections for positrons of energies 2–400 eV in He, Ne, Ar, and Kr. *J. Phys. B Atomic Mol. Phys.* **1973**, *6*, L201–L203. [CrossRef]
18. Williams, J.F.; Willis, B.A. The scattering of electrons from inert gases. I. Absolute differential elastic cross sections for argon atoms. *J. Phys. B.* **1975** *8*, 1670. [CrossRef]
19. Bromberg, J.P. Absolute differential cross sections of electrons elastically scattered by the rare gases. I. Small angle scattering between 200 and 700 eV. *J. Chem. Phys.* **1974**, *61*, 963. [CrossRef]
20. Gupta, S.C.; Rees, J.A. Absolute differential cross sections for 100 eV electrons elastically scattered by helium, neon, and argon. *J. Phys. B* **1975**, *8*, 1267. [CrossRef]
21. Kauppila, W.E.; Stein, T.S.; Jesion, G. Direct Observation of a Ramsauer-Townsend Effect in Positron-Argon Collisions. *Phys. Rev. Lett.* **1976**, *36*, 580. [CrossRef]
22. Dubois, R.D.; Rudd, M.E. Differential cross sections for elastic scattering of electrons from argon, neon, nitrogen and carbon monoxide. *J. Phys. B* **1976**, *9*, 2657. [CrossRef]
23. Vuskovic, L.; Kurepa, M.V. Differential cross sections of 60–150 eV electrons elastically scattered in argon. *J. Phys. B* **1976**, *9*, 837. [CrossRef]
24. Jansen, R.H.J.; de Heer, F.J.; Luyken, H.J.; van Wingerden, B.; Blaauw, H.J. Absolute differential cross sections for elastic scattering of electrons by helium, neon, argon and molecular nitrogen. *J. Phys. B* **1976**, *9*, 185. [CrossRef]
25. Kauppila, W.E.; Stein, T.S.; Smart, J.H, Dababneh, M.S.; Ho Y.K.; Downing, J.P.; Pol, V. Measurements of total scattering cross sections for intermediate-energy positrons and electrons colliding with helium, neon, and argon. *Phys. Rev. A* **1981** *24*, 725–742. [CrossRef]
26. Jones, A.C.L.; Makochekanwa, C.; Caradonna, P.; Slaughter, D.S.; Machacek, J.R.; McEachran, R.P.; Sullivan, J.P.; Buckman, S.J.; Stauffer, A.D.; Bray, I.; et al. Positron scattering from neon and argon. *Phys. Rev. A* **2011**, *83*, 032701. [CrossRef]
27. Hyder, G.M.A.; Dababneh, M.S.; Hsieh, Y.-F.; Kauppila, W.E.; Kwan, C.K.; Mahdavi-Hezaveh, M.; Stein, T.S. Positron Differential Elastic-Scattering Cross-Section Measurements for Argon. *Phys. Rev. Lett.* **1986**, *57*, 2252. [CrossRef]
28. Griffith, T.C.; Heyland, G.R. Experimental aspects of the study of the interaction of low-energy positrons with gases. *Phys. Rep* **1978**, *39C*, 169–277. [CrossRef]
29. Tsai, J.-S.; Lebnow, L.; Paul, D.A.L. Measurement of total cross sections (e+, Ne) and (e+, Ar). *Can. J. Phys.* **1976**, *54*, 1741. [CrossRef]
30. Kauppila, W.E.; Kwan, C.K.; Stein, T.S.; Zhou, S. i Evidence for channel-coupling effects in positron scattering by sodium and potassium atoms. *J. Phys. B Atomic Mol. Opt. Phys.* **1994**, *27*, L551–L555. [CrossRef]
31. Kwan, C.K.; Kauppila. W.E.; Lukaszew. R.A.; Parikh. S.P.; Stein, T.S.; Wan, Y.J.; Dababneh, M.S. Total crosssection measurements for positrons and electrons scattered by sodium and potassium atoms. *Phys. Rev.* **1991**, *44*, 1620. [CrossRef] [PubMed]

32. Stein, T.S.; Jiang, J.; Kauppila, W.E.; Kwan, C.K.; Li, H.; Surdutovich, A.; Zhou, S. Measurements of total and (or) positronium-formation cross sections for positrons scattered by alkali, magnesium, and hydrogen atoms. *Can. J. Phys.* **1996**, *74*, 313–333. [CrossRef]
33. Stein, T.; Harte, M.; Jiang, J.; Kauppila, W.; Kwan, C.; Li, H.; Zhou, S. Measurements of positron scattering by hydrogen, alkali metal, and other atoms. *Nucl. Instr. Meth. Phys. Res. B* **1998**, *143*, 68–80. [CrossRef]
34. Ferch, J.; Granitza, B.; Masche, C.; Raith, W. Electron-argon total cross section measurements at low energies by time-of-flight spectroscopy. *J. Phys. B* **1985**, *18*, 967. [CrossRef]
35. Williams, J.F. A phaseshift analysis of experimental angular distributions of electrons elastically scattered from He, Ne and Ar over the range 0.5 to 20 eV. *J. Phys. B* **1979**, *12*, 265. [CrossRef]
36. Charlton, M.; Grifith, T.C.; Heyland, G.R.; Twomey, T.R. Total scattering cross sections for low-energy electrons in helium and argon. *J. Phys. B* **1980**, *13*, L239. [CrossRef]
37. Jost, K.; Bisling, P.G.F.; Eschen, F.; Felsmann, M.; Walther, L. Abstracts of Contributed Papers. In Proceedings of the Thirteenth International Conference on the Physics of Electronic and Atomic Collisions, Berlin, Germany, 27 July–2 August 1983; Eichler, J., Fritsch, W., Hertel, I.V., Stolterfoin, N., Wille, W., Eds.; North-Holland: Amsterdam, The Netherlands, 1983.
38. Wagenaar, R.W.; de Heer, F.J. Total cross sections for electron scattering from Ar, Kr and Xe. *J. Phys. B* **1985**, *18*, 2021. [CrossRef]
39. Nicket, J.C.; Imre, K.; Register, D.F.; Trajmar, S. Total electron scattering cross sections. I. He, Ne, Ar, Xe. *J. Phys. B* **1985**, *18*, 125. [CrossRef]
40. Mehr, J. Winkelverteilungen elastisch an Edelgas-Atomstrahlen gestreuter Elektronen; Spinpolarisation eines an Argon gestreuten 40 eV-Elektronenstrahls. *Z. Phys.* **1967**, *198*, 345. [CrossRef]
41. Beerlage, M.J.M.; Qing, Z.; Van der Wiel, M.J. The polarisation of electrons elastically scattered from argon and krypton at energies between 10 and 50 eV. *J. Phys. B* **1981**, *14*, 4627. [CrossRef]
42. Schackert, K. Spin polarization of slow electrons elastically scattered by noble gas atomic beams. *Z. Phys.* **1968**, *213*, 316. [CrossRef]
43. Ajmera, M.P.; Chung, K.T. Simplified variational method for P-wave electron-hydrogen scattering. *Phys. Rev. A* **1974**, *10*, 1013. [CrossRef]
44. Drachman, R.J. Positron-Hydrogen Scattering at Low Energies. *Phys. Rev.* **1965** *138*, A1582. [CrossRef]
45. Bhatia, A.K. Positron Impact Excitation of the 2S State of Atomic Hydrogen. *Atoms* **2019**, *7*, 69. [CrossRef]
46. Bhatia, A.K. Hybrid theory of P-wave electron-Li2+ elastic scattering and photoabsorption in two-electron systems. *Phys. Rev. A* **1983**, *87*, 042705. [CrossRef]
47. Bhatia, A.K. 2006 Electron-wave elastic scattering and photoabsorption in two-electron systems. *Phys. Rev. A* **2007**, *73*, 012705. [CrossRef]
48. Bhatia, A.K. Hybrid theory of electron-hydrogen elastic scattering. *Phys. Rev. A* **2007**, *75*, 032713. [CrossRef]
49. Joshipura, K.N.; Vinodkumar, M.; Limbachiya, C.G.; Antony, B.K. Calculated total cross sections of electron-impact ionization and excitations in tetrahedral (XY_4) and SF_6 molecules. *Phys. Rev. A* **2004**, *69*, 022705. [CrossRef]
50. Sinha, N.; Patel, D.; Antony, B. Positron Scattering: Total Elastic and Grand Total Cross Sections for Molecules of Astrophysical Importance. *ChemistrySelect* **2019**, *4*, 4575–4581. [CrossRef]
51. Sloan, I.H. The method of polarized orbitals for the elastic scattering of slow electrons by ionized helium and atomic hydrogen. *Proc. R. Soc. Lond.* **1964**, *281*, 151–163.
52. Zhou, Y.; Ratnavelu, K.; McCarthy, I.E. Momentum-space coupled-channel optical method for positron-hydrogen scattering. *Phys. Rev. A* **2005**, *71*, 042703. [CrossRef]
53. Dalgarno, A. Perturbation theory for atomic systems. *Proc. R. Soc. A* **1959**, *251*, 282–90. [CrossRef]
54. Kaneko, S. Electrical Polarizabilities of Rare Gas Atoms by the Hartree-Fock Wave Functions. *J. Phys. Soc. Jpn.* **1959**, *14*, 1600–1611. [CrossRef]
55. McEachran, R.P.; Stauffer, A.D. Positron scattering from helium. *J. Phys. B Atomic Mol. Opt. Phys.* **2019**, *52*, 115203. [CrossRef]
56. Feshbach, H. A unified theory of nuclear reactions. II. *Ann. Phys* **1962** *19*, 287. [CrossRef]
57. Brown, C.J.; Humberston, J.W. Positronium formation in positron-hydrogen scattering. *J. Phys. B Atomic Mol. Opt. Phys.* **1985**, *18*, L401. [CrossRef]
58. Dzuba, V.A.; Flambaum, V.V.; Gribakin, G.F.; King, W.A. Many-body calculations of positron scattering and annihilation from noble-gas atoms. *J. Phys. B Atomic Mol. Opt. Phys.* **1996**, *29*, 3151. [CrossRef]

59. Green, D.G.; Ludlow, J.A.; Gribakin, G.F. Positron scattering and annihilation on noble-gas atoms. *Phys. Rev. A* **2014**, *90*, 032712. [CrossRef]
60. Germano, J.S.E.; Lima, M.A.P. Schwinger multichannel method for positron-molecule scattering. *Phys. Rev. A* **1993**, *47*, 3976. [CrossRef]
61. Arretche, F.; da Costa, R.F.; Sanchez, S.D.A.; Hisi, A.N.S.; De Oliveira, E.M.; Varella, M.D.N.; Lima, M.A.P. Similarities and differences in $e^{\pm}$–molecule scattering: Applications of the Schwinger multichannel method. *Nucl. Instrum. Methods Phys. Res. B* **2006**, *247*, 13–19. [CrossRef]
62. Barbosa, A.S.; Bettega, M.H. Calculated cross sections for elastic scattering of slow positrons by silane. *Phys. Rev. A* **2017**, *96*, 042715. [CrossRef]
63. Higgins, K.; Burke, P.G.; Walters, H.R. Positron scattering by atomic hydrogen at intermediate energies. *J. Phys. B Atomic Mol. Opt. Phys.* **1990**, *23*, 1345. [CrossRef]
64. Scholz, T.; Scott, P.; Burke, P.G. Electron-hydrogen-atom scattering at intermediate energies. *J. Phys. B* **1988**, *21*, L139–L145 [CrossRef]
65. Grant, I.P.; McKenzie, B.J.; Norrington, P.H.; Mayers, D.F.; Pyper, N.C. An atomic multiconfigurational Dirac-Fock package. *Comput. Phys. Commun.* **1980**, *21*, 207–231. [CrossRef]
66. McAlinden, M.T.; Kernoghan, A.A.; Walters, H.R.J. Cross-channel coupling in positron-atom scattering. *Hyperfine Interact.* **1994** *89*, 161–194. [CrossRef]
67. Mukherjee, T.; Mukherjee, M. Low-energy positron–nitrogen-molecule scattering: A rovibrational close-coupling study. *Phys. Rev. A* **2015**, *91*, 062706. [CrossRef]
68. Jain, A.; Baluja, K.L. Total (elastic plus inelastic) cross sections for electron scattering from diatomic and polyatomic molecules at 10–5000 eV: H_2, Li_2, HF, CH_4, N_2, CO, C_2H_2, HCN, O_2, HCl, H_2S, PH_3, SiH_4, and CO_2. *Phys. Rev. A* **1992**, *45*, 202–218. [CrossRef]
69. Gao, J.; Peacher, J.L.; Madison, D.H. An elementary method for calculating orientation-averaged fully differential electron-impact ionization cross sections for molecules. *J. Chem. Phys.* **2005**, *123*, 204302. [CrossRef]
70. Madison, D.H.; Al-Hagan, O. The distorted-wave Born approach for calculating electron-impact ionization of molecules. *J. Atom. Mol. Opt. Phys.* **2010**, *2010*, 367180. [CrossRef]
71. Singh, S.; Dutta, S.; Naghma, R.; Antony, B. Theoretical formalism to estimate the positron scattering cross section. *J. Phys. Chem. A* **2016**, *120*, 5685–5692. [CrossRef]
72. Raizada, R.; Baluja, K.L. Total cross sections of positron scattering from various molecules using the rule of additivity. *Phys. Rev. A* **1997**, *55*, 1533–1536. [CrossRef]
73. Morrison, M.A.; Gibson, T.L.; Austin, D. Polarisation potentials for positron-molecule collisions: Positron-H_2 scattering. *J. Phys. B Atomic Mol. Opt. Phys.* **1984**, *17*, 2725. [CrossRef]
74. Raj, D. Total cross sections for positron scattering by molecules. *Phys. Lett. A* **1993**, *174*, 304–307. [CrossRef]
75. Khandker, M.H.; Arony, N.T.; Haque, A.K.F.; Maaza, M.; Billah, M.M.; Uddin, M.A. Scattering of $e^{\pm}$ from N_2 in the energy range 1 eV–10 keV. *Mol. Phys.* **2019**, 1–15. [CrossRef]
76. Horbatsch, M.; Darewych, J.W. Model potential description of low-energy e^+-CO_2 scattering. *J. Phys. B Atomic Mol. Opt. Phys.* **1983**, *16*, 4059. [CrossRef]
77. Raj, D.; Tomar, S. Scattering of positrons by hydrocarbons at intermediate energies. *Indian J. Phys* **1996**, *70B*, 375–383.
78. Blanco, F.; Ellis-Gibbings, L.; García, G. Screening corrections for the interference contributions to the electron and positron scattering cross sections from polyatomic molecules. *Chem. Phys. Lett.* **2016** *645*, 71–75. [CrossRef]
79. Srivastava, M.K.; Pathak, A. Total scattering cross sections for intermediate-and high-energy positrons in H_2. *J. Phys. B Atomic Mol. Opt. Phys.* **1981**, *14*, L579. [CrossRef]
80. Ellis-Gibbings, L.; Blanco, F.; García, G. Positron interactions with nitrogen and oxygen molecules: Elastic, inelastic and total cross sections. *Eur. Phys. J. D* **2019**, *73*, 266. [CrossRef]
81. Chiari, L.; Zecca, A.; Girardi, S.; Trainotti, E.; Garcia, G.; Blanco, F.; McEachran, R.P.; Brunger, M.J. Positron scattering from O_2. *J. Phys. B Atomic Mol. Opt. Phys.* **2012** *45*, 215206. [CrossRef]
82. Blanco, F.; Roldán, A.M.; Krupa, K.; McEachran, R.P.; White, R.D.; Marjanović, S.; Petrović, Z.L.; Brunger, M.J.; Machacek, J.R.; Buckman, S.J.; et al. Scattering data for modelling positron tracks in gaseous and liquid water. *J. Phys. B Atomic Mol. Opt. Phys.* **2016**, *49*, 145001. [CrossRef]

83. Tattersall, W.; Chiari, L.; Machacek, J.R.; Erson, E.; White, R.D.; Brunger, M.J.; Buckman, S.J.; Garcia, G.; Blanco, F.; Sullivan, J.P. Positron interactions with water–total elastic, total inelastic, and elastic differential cross section measurements. *J. Chem. Phys.* **2014**, *140*, 044320. [CrossRef]
84. Chiari, L.; Zecca, A.; Trainotti, E.; Garcia, G.; Blanco, F.; Bettega, M.H.; Sanchez, S.D.A.; Varella, M.T.D.N.; Lima, M.A.; Brunger, M.J. Positron and electron collisions with nitrous oxide: Measured and calculated cross sections. *Phys. Rev. A* **2013** , *88*, 022708. [CrossRef]
85. Chiari, L.; Zecca, A.; García, G.; Blanco, F.; Brunger, M.J. Low-energy positron and electron scattering from nitrogen dioxide. *J. Phys. B Atomic Mol. Opt. Phys.* **2013** *46*, 235202. [CrossRef]
86. Zecca, A.; Trainotti, E.; Chiari, L.; García, G.; Blanco, F.; Bettega, M.H.F.; do N Varella, M.T.; Lima, M.A.P.; Brunger, M.J. An experimental and theoretical investigation into positron and electron scattering from formaldehyde. *J. Phys. B Atomic Mol. Opt. Phys.* **2011**, *44*, 195202. [CrossRef]
87. Chiari, L.; Palihawadana, P.; Machacek, J.R.; Makochekanwa, C.; García, G.; Blanco, F.; McEachran, R.P.; Brunger, M.J.; Buckman, S.J.; Sullivan, J.P. Experimental and theoretical cross sections for positron collisions with 3-hydroxy-tetrahydrofuran. *J. Chem. Phys.* **2013**, *138*, 074302. [CrossRef] [PubMed]
88. Chiari, L.; Zecca, A.; Blanco, F.; García, G.; Brunger, M.J. Cross sections for positron and electron collisions with an analog of the purine nucleobases: Indole. *Phys. Rev. A* **2015** *91*, 012711. [CrossRef]
89. Anderson, E.K.; Boadle, R.A.; Machacek, J.R.; Chiari, L.; Makochekanwa, C.; Buckman, S.J.; Brunger, M.J.; Garcia, G.; Blanco, F.; Ingolfsson, O.; et al. Low energy positron interactions with uracil—Total scattering, positronium formation, and differential elastic scattering cross sections. *J. Chem. Phys.* **2014** *141*, 034306. [CrossRef]
90. Chiari, L.; Zecca, A.; Blanco, F.; García, G.; Brunger, M.J. Experimental and theoretical cross sections for positron scattering from the pentane isomers. *J. Chem. Phys.* **2016** *144*, 084301. [CrossRef]
91. Chiari, L.; Zecca, A.; Blanco, F.; García, G.; Perkins, M.V.; Buckman, S.J.; Brunger, M.J. Cross sections for positron impact with 2, 2, 4-trimethylpentane. *J. Phys. Chem. A* **2014**, *118*, 6466–6472. [CrossRef]
92. Chiari, L.; Zecca, A.; Blanco, F.; García, G.; Brunger, M.J. Positron scattering from vinyl acetate. *J. Phys. B Atomic Mol. Opt. Phys.* **2014**, *47*, 175202. [CrossRef]
93. Sanz, A.G.; Fuss, M.C.; Blanco, F.; Mašín, Z.; Gorfinkiel, J.D.; McEachran, R.P.; Brunger, M.J.; García, G. Cross-section calculations for positron scattering from pyrimidine over an energy range from 0.1 to 10000 eV. *Phys. Rev. A* **2013**, *88*, 062704. [CrossRef]
94. Zecca, A.; Chiari, L.; García, G.; Blanco, F.; Trainotti, E.; Brunger, M.J. Total cross sections for positron and electron scattering from pyrimidine. *J. Phys. B Atomic Mol. Opt. Phys.* **2010**, *43*, 215204. [CrossRef]
95. Stevens, D.; Babij, T.J.; Machacek, J.R.; Buckman, S.J.; Brunger, M.J.; White, R.D.; García, G.; Blanco, F.; Ellis-Gibbings, L.; Sullivan, J.P. Positron scattering from pyridine. *J. Chem. Phys.* **2018**, *148*, 144308. [CrossRef] [PubMed]
96. Zecca, A.; Trainotti, E.; Chiari, L.; Bettega, M.H.F.; Sanchez, S.D.A.; Varella, M.D.N.; Lima, M.A.P.; Brunger, M.J. Positron scattering from the cyclic ethers oxirane, 1, 4-dioxane, and tetrahydropyran. *J. Chem. Phys.* **2012**, *136*, 124305. [CrossRef] [PubMed]
97. Zecca, A.; Chiari, L.; García, G.; Blanco, F.; Trainotti, E.; Brunger, M.J. Total cross-sections for positron and electron scattering from α-tetrahydrofurfuryl alcohol. *New J. Phys.* **2011**, *13*, 063019. [CrossRef]
98. Humberston, J.W.; Van Reeth, P.; Watts, M.S.T.; Meyerhof, W.E. Positron-hydrogen scattering in the vicinity of the positronium formation threshold. *J. Phys. B Atomic Mol. Opt. Phys.* **1997**, *30*, 2477. [CrossRef]
99. Van Reeth, P.; Humberston, J.W. Theoretical studies of threshold features in the cross-sections for low-energy e^+–H and e^+–He scattering. *Nucl. Instrum. Methods Phys. Res. B* **2000**, *171*, 106–112. [CrossRef]
100. Armour, E.A.G.; Baker, D.J.; Plummer, M. The theoretical treatment of low-energy e^+-H_2 scattering using the kohn variational method. *J. Phys. B Atomic Mol. Opt. Phys.* **1990**, *23*, 3057. [CrossRef]
101. Weiss, L.I.; Pinho, A.S.; Michelin, S.E.; Fujimoto, M.M. Electronic excitation cross section in positron scattering by H_2 molecules using distorted-wave method. *Eur. Phys. J. D* **2018**, *72*, 35. [CrossRef]
102. Basu, M.; Mazumdar, P.S.; Ghosh, A.S. Ionisation cross sections in positron-helium scattering. *J. Phys. B Atomic Mol. Opt. Phys.*, **1985**, *18*, 369. [CrossRef]
103. Parcell, L.A.; McEachran, R.P.; Stauffer, A.D. Positron excitation of the 21S state of helium. *J. Phys. B Atomic Mol. Opt. Phys.* **1983**, *16*, 4249. [CrossRef]
104. Khan, P.; Ghosh, A.S. Positronium formation in positron-helium scattering. *Phys. Rev. A* **1983**, *28*, 2181–2189. [CrossRef]

105. Parcell, L.A. McEachran, R.P.; Stauffer, A.D. Positron scattering from neon. *Nucl. Instrum. Methods Phys. Res. B* **1998**, *143*, 37–40. [CrossRef]
106. Parcell, L.A.; McEachran, R.P.; Stauffer, A.D. Positron scattering from argon and krypton. *Nucl. Instrum. Methods Phys. Res. B* **2000**, *171*, 113–118. [CrossRef]
107. Gribakin, G.F.; Ludlow, J. Many-body theory of positron-atom interactions. *Phys. Rev. A* **2004**, *70*, 032720. [CrossRef]
108. Gribakin, G.F.; King, W.A. Positron scattering from Mg atoms. *Can. J. Phys.* **1996**, *74*, 449–459. [CrossRef]
109. Meyerhof, W.E.; Laricchia, G.; Moxom, J.; Humberston, J.W.; Watts, M.S.T. Positron scattering on atoms and molecules near the positronium threshold. *Can. J. Phys.* **1996**, *74*, 427–433. [CrossRef]
110. Bartschat, K. Direct ionization of heavy noble gases by positron impact. *Phys. Rev. A* **2005**, *71*, 032718. [CrossRef]
111. Moxom, J.; Laricchia, G.; Charlton, M.; Kovar, A.; Meyerhof, W.E. Threshold effects in positron scattering on noble gases. *Phys. Rev. A* **1994**, *50*, 3129–3133. [CrossRef]
112. Danby, G.; Tennyson, J. Positron-HF collisions: Prediction of a weakly bound state. *Phys. Rev. Lett.* **1988**, *61*, 2737. [CrossRef] [PubMed]
113. Tennyson, J. Low-energy, elastic positron-molecule collisions using the R-matrix method: e^+-He_2 and e^+-N_2. *J. Phys. B Atomic Mol. Opt. Phys.* **1986**, *19*, 4255. [CrossRef]
114. Danby, G.; Tennyson, J. Differential cross sections for elastic positron-H2 collisions using the R-matrix method. *J. Phys. B Atomic Mol. Opt. Phys.* **1990**, *23*, 1005. [CrossRef]
115. Zhang, R.; Baluja, K.L.; Franz, J.; Tennyson, J. Positron collisions with molecular hydrogen: Cross sections and annihilation parameters calculated using the R-matrix with pseudo-states method. *J. Phys. B Atomic Mol. Opt. Phys.* **2011**, *44*, 035203. [CrossRef]
116. Baluja, K.L.; Zhang, R.; Franz, J.; Tennyson, J. Low-energy positron collisions with water: Elastic and rotationally inelastic scattering. *J. Phys. B Atomic Mol. Opt. Phys.* **2007**, *40*, 3515. [CrossRef]
117. Franz, J.; Baluja, K.L.; Zhang, R.; Tennyson, J. Polarisation effects in low-energy positron–molecule scattering. *Nucl. Instrum. Methods Phys. Res. B* **2008**, *266*, 419–424. [CrossRef]
118. Zhang, R.; Galiatsatos, P.G.; Tennyson, J. Positron collisions with acetylene calculated using the R-matrix with pseudo-states method. *J. Phys. B Atomic Mol. Opt. Phys.* **2011**, *44*, 195203. [CrossRef]
119. Lino, J.L. Improving the wavefunction of the Schwinger multichannel method: Application for positron elastic scattering by He atom. *Phys. Scr.* **2007**, *76*, 521. [CrossRef]
120. Sanchez, S.D.A.; Lima, M.A.P. The influence of f-type function in positron-He/positron-H2 scattering with the Schwinger multichannel method. *Nucl. Instrum. Methods Phys. Res. B* **2008**, *266*, 447–451. [CrossRef]
121. Zanin, G.L.; Tenfen, W.; Arretche, F. Rotational excitation of H 2 by positron impact in adiabatic rotational approximation. *Eur. Phys. J. D* **2016**, *70*, 179. [CrossRef]
122. Varella, M.T.D.N.; Lima, M.A. Near-threshold vibrational excitation of H_2 by positron impact: A projection-operator approach. *Phys. Rev. A* **2007**, *76*, 052701. [CrossRef]
123. Lino, J.L. Improving the wavefunction of the Schwinger multichannel method for positron scattering—II: Application for elastic and inelastic e^+-H_2 scattering. *Phys. Scr.* **2009**, *80*, 065303. [CrossRef]
124. Lino, J.L.; Germano, J.S.E.; Lima, M.A.P. Electronic excitation of H2 by positron impact: An application of the Schwinger multichannel method. *J. Phys. B Atomic Mol. Opt. Phys.* **1994**, *27*, 1881. [CrossRef]
125. Lino, J.L.S. Electronic excitation of H_2, CO, and N_2 by positron impact. *Rev. Mex. Fis.* **2017**, *63*, 303–307.
126. Lino, J.L.; Germano, J.S.; da Silva, E.P.; Lima, M.A. Elastic cross sections and annihilation parameter for e^+-H_2 scattering using the Schwinger multichannel method. *Phys. Rev. A* **1998**, *58*, 3502. [CrossRef]
127. Seidel, E.P.; Barp, M.V.; Tenfen, W.; Arretche, F. Elastic scattering and rotational excitation of Li_2 by positron impact. *J. Electron. Spectros. Relat. Phenomena* **2018**, *227*, 9–14. [CrossRef]
128. de Carvalho, C.R.; do N Varella, M.T.; Lima, M.A.; da Silva, E.P.; Germano, J.S. Progress with the Schwinger multichannel method in positron–molecule scattering. *Nucl. Instrum. Methods Phys. Res. B* **2000**, *171*, 33–46. [CrossRef]
129. Mazon, K.T.; Tenfen, W.; Michelin, S.E.; Arretche, F.; Lee, M.T.; Fujimoto, M.M. Vibrational cross sections for positron scattering by nitrogen molecules. *Phys. Rev. A* **2010**, *82*, 032704. [CrossRef]
130. Chaudhuri, P.; Varella, M.T.D.N.; de Carvalho, C.R.; Lima, M.A. Electronic excitation of N_2 by positron impact. *Phys. Rev. A* **2004**, *69*, 042703. [CrossRef]

131. Chaudhuri, P.; Varella, M.T.D.N.; de Carvalho, C.R.; Lima, M.A. Positron impact electronic excitation of N_2. *Nucl. Instrum. Methods Phys. Res. B* **2004**, *221*, 69–75. [CrossRef]
132. da Silva, E.P.; Varella, M.T.D.N.; Lima, M.A. Electronic excitation of CO by positron impact. *Phys. Rev. A* **2005**, *72*, 062715. [CrossRef]
133. Sanchez, S.D.A.; Arretche, F.; do N Varella, M.T.; Lima, M.A.P. Low energy positron scattering by SF6 and CO_2. *Phys. Scr.* **2004**, *T110*, 276. [CrossRef]
134. Sanchez, S.D.A.; Arretche, F.; Lima, M.A.P. Low-energy positron scattering by CO_2. *Phys. Rev. A* **2008**, *77*, 054703. [CrossRef]
135. Arretche, F.; Tenfen, W.; Mazon, K.T.; Michelin, S.E.; Lima, M.A.P.; Lee, M.T.; Machado, L.E.; Fujimoto, M.M.; Pessoa, O.A. Low energy scattering of positrons by H_2O. *Nucl. Instrum. Methods Phys. Res. B* **2010**, *268*, 178–182. [CrossRef]
136. Arretche, F.; Barp, M.V.; Seidel, E.P.; Tenfen, W. Electronic excitation of H_2O by positron impact. *Eur. Phys. J. D* **2020**, *74*, 1–7. [CrossRef]
137. Lino, J.L. Elastic scattering of positrons by H_2O at low energies. *Rev. Mex. Fis.* **2014**, *60*, 156–160.
138. Zecca, A.; Chiari, L.; Trainotti, E.; Sarkar, A.; Sanchez, S.D.A.; Bettega, M.H.F.; Varella, M.D.N.; Lima, M.A.P.; Brunger, M.J. Positron scattering from methane. *Phys. Rev. A* **2012**, *85*, 012707. [CrossRef]
139. Zecca, A.; Chiari, L.; Sarkar, A.; Lima, M.A.; Bettega, M.H.; Nixon, K.L.; Brunger, M.J. Positron scattering from formic acid. *Phys. Rev. A* **2008**, *78*, 042707. [CrossRef]
140. Barbosa, A.S.; Blanco, F.; García, G.; Bettega, M.H. Theoretical study on positron scattering by benzene over a broad energy range. *Phys. Rev. A* **2019**, *100*, 042705. [CrossRef]
141. Barbosa, A.S.; Pastega, D.F.; Bettega, M.H. Low-energy positron scattering by pyrimidine. *J. Chem. Phys.* **2015**, *143*, 244316. [CrossRef]
142. Barbosa, A.S.; Sanchez, S.D.A.; Bettega, M.H. Bound state in positron scattering by allene. *Phys. Rev. A* **2017**, *96*, 062706. [CrossRef]
143. Barbosa, A.S.; Bettega, M.H. Elastic collisions of low-energy positrons with tetrahydrofuran. *J. Chem. Phys.* **2019**, *150*, 184305. [CrossRef] [PubMed]
144. Bettega, M.H.; Sanchez, S.D.A.; Varella, M.T.D.N.; Lima, M.A.; Chiari, L.; Zecca, A.; Trainotti, E.; Brunger, M.J. Positron collisions with ethene. *Phys. Rev. A* **2012**, *86*, 022709. [CrossRef]
145. de Carvalho, C.R.; Varella, M.T.D.N.; Lima, M.A.; da Silva, E.P. Elastic positron scattering by C_2H_2: Differential cross sections and virtual state formation. *Phys. Rev. A* **2003**, *68*, 062706. [CrossRef]
146. Chiari, L.; Zecca, A.; Trainotti, E.; Bettega, M.H.F.; Sanchez, S.D.A.; Varella, M.D.N.; Lima, M.A.P.; Brunger, M.J. Cross sections for positron scattering from ethane. *Phys. Rev. A* **2013**, *87*, 032707. [CrossRef]
147. Makochekanwa, C.; Kato, H.; Hoshino, M.; Bettega, M.H.F.; Lima, M.A.P.; Sueoka, O.; Tanaka, H. Electron and positron scattering from 1, 1-$C_2H_2F_2$. *J. Chem. Phys.* **2007**, *126*, 164309. [CrossRef]
148. Moreira, G.M.; Bettega, M.H. Elastic Scattering of Slow Positrons by Pyrazine. *J. Phys. Chem. A* **2019**, *123*, 9132–9136. [CrossRef]
149. Moreira, G.M.; Bettega, M.H. Low-energy positron collisions with CH_2O.......H_2O complexes. *Eur. Phys. J. D* **2017**, *71*, 1–5. [CrossRef]
150. Nunes, F.B.; Bettega, M.H.; Sanchez, S.D. A. Positron collisions with C3H6 isomers. *J. Phys. B Atomic Mol. Opt. Phys.* **2015**, *48*, 165201. [CrossRef]
151. Nunes, F.B.; Bettega, M.H.; Sanchez, S.D.A. Positron and electron scattering by glycine and alanine: Shape resonances and methylation effect. *J. Chem. Phys.* **2016**, *145*, 214313. [CrossRef]
152. da Silva, E.P.; Germano, J.S.; Lima, M.A. Annihilation Dynamics of Positrons in Molecular Environments: Theoretical Study of Low-Energy Positron-C_2H_4 Scattering. *Phys. Rev. Lett.* **1996**, *77*, 1028. [CrossRef]
153. Silva, F.M.; Bettega, M.H.; Sanchez, S.D.A. Low-energy positron and electron scattering by methylamine. *Eur. Phys. J. D* **2014**, *68*, 12. [CrossRef]
154. do N Varella, M.T.; Sanchez, S.D.A.; Bettega, M.H.; Lima, M.A.; Chiari, L.; Zecca, A.; Trainotti, E.; Brunger, M.J. Low-energy positron scattering from iodomethane. *J. Phys. B Atomic Mol. Opt. Phys.* **2013**, *46*, 175202. [CrossRef]
155. Ghosh, A.S.; Mukherjee, T.; Biswas, P.K.; Jain, A. Positron-CO collisions using rotational LFCCA. *J. Phys. B Atomic Mol. Opt. Phys.* **1993**, *26*, L23. [CrossRef]
156. Hewitt, R.N.; Noble, C.J.; Bransden, B.H. Positronium formation in positron-hydrogen scattering. *J. Phys. B Atomic Mol. Opt. Phys.* **1990**, *23*, 4185. [CrossRef]

157. Mitroy, J.; Ratnavelu, K. The positron-hydrogen system at low energies. *J. Phys. B Atomic Mol. Opt. Phys.* **1995**, *28*, 287. [CrossRef]
158. Mitroy, J. An calculation of positron-hydrogen scattering at intermediate energies. *J. Phys. B Atomic Mol. Opt. Phys.* **1996**, *29*, L263. [CrossRef]
159. Wakid, S.E.A.; LaBahn, R.W. Positronium formation in positron-hydrogen collisions. *Phys. Rev. A* **1972**, *6*, 2039. [CrossRef]
160. Zhou, Y.; Lin, C.D. Hyperspherical close-coupling calculation of positronium formation cross sections in positron-hydrogen scattering at low energies. *J. Phys. B Atomic Mol. Opt. Phys.* **1994**, *27*, 5065. [CrossRef]
161. Kadyrov, A.S.; Bray, I. Two-center convergent close-coupling approach to positron-hydrogen collisions. *Phys. Rev. A* **2002**, *66*, 012710. [CrossRef]
162. Ryzhikh, G.; Mitroy, J. Positron scattering from atomic sodium. *J. Phys. B Atomic Mol. Opt. Phys.* **1997**, *30*, 5545. [CrossRef]
163. Ward, S.J.; Horbatsch, M.; McEachran, R.P.; Stauffer, A.D. Close-coupling approach to positron scattering for lithium, sodium and potassium. *J. Phys. B Atomic Mol. Opt. Phys.* **1989**, *22*, 1845. [CrossRef]
164. Ward, S.J.; Horbatsch, M.; McEachran, R.P.; Stauffer, A.D. Close-coupling approach to positron scattering from potassium. *J. Phys. B Atomic Mol. Opt. Phys.* **1988**, *21*, L611. [CrossRef]
165. Khan, P.; Dutta, S.; Ghosh, A. Positron-lithium scattering using the eigenstate expansion method. *J. Phys. B Atomic Mol. Opt. Phys.* **1987**, *20*, 2927. [CrossRef]
166. Natchimuthu, N.; Ratnavelu, K. Optical potential study of positron scattering by atomic sodium at intermediate energies. *Phys. Rev. A* **2001** *63*, 052707. [CrossRef]
167. McCarthy, I.E.; Ratnavelu, K.; Zhou, Y. Calculation of total cross sections for electron and positron scattering on sodium and potassium. *J. Phys. B Atomic Mol. Opt. Phys.* **1993**, *26*, 2733. [CrossRef]
168. McAlinden, M.T.; Kernoghan, A.A.; Walters, H.R.J. Positron scattering by potassium. *J. Phys. B Atomic Mol. Opt. Phys.* **1996**, *29*, 555. [CrossRef]
169. McAlinden, M.T.; Kernoghan, A.A.; Walters, H.R.J. Positron scattering by lithium. *J. Phys. B Atomic Mol. Opt. Phys.* **1997**, *30*, 1543. [CrossRef]
170. Kernoghan, A.A.; McAlinden, M.T.; Walters, H.R.J. Positron scattering by rubidium and caesium. *J. Phys. B Atomic Mol. Opt. Phys.* **1996**, *29*, 3971. [CrossRef]
171. Mitroy, J.; Ratnavelu, K. Close coupling theory of positron scattering from alkali atoms. *Aust. J. Phys.* **1994**, *47*, 721–742. [CrossRef]
172. Hewitt, R.N.; Joachain, C.J.; Noble, C.J.; Bransden, B.H. Coupled-channel calculations of e^+–Mg scattering. *Can. J. Phys.* **1996**, *74*, 559–563. [CrossRef]
173. Utamuratov, R.; Fursa, D.V.; Kadyrov, A.S.; Lugovskoy, A.V.; Savage, J.S.; Bray, I. Two-center convergent-close-coupling calculations of positron scattering on magnesium. *Phys. Rev. A* **2012**, *86*, 062702. [CrossRef]
174. Fursa, D.V.; Bray, I. Convergent close-coupling method for positron scattering from noble gases. *New J. Phys.* **2012**, *14*, 035002. [CrossRef]
175. McEachran, R.P.; Horbatsch, M.; Stauffer, A.D. Positron scattering from rubidium. *J. Phys. B Atomic Mol. Opt. Phys.* **1991**, *24*, 1107. [CrossRef]
176. Utamuratov, R.; Fursa, D.V.; Mori, N.; Kadyrov, A.S.; Bray, I.; Zammit, M.C. Positron-impact electronic excitations and mass stopping power of H_2. *Phys. Rev. A* **2019**, *99*, 042705. [CrossRef]
177. Zammit, M.C.; Fursa, D.V.; Savage, J.S.; Bray, I.; Chiari, L.; Zecca, A.; Brunger, M.J. Adiabatic-nuclei calculations of positron scattering from molecular hydrogen. *Phys. Rev. A* **2017**, *95*, 022707. [CrossRef]
178. Zammit, M.C.; Fursa, D.V.; Bray, I. Convergent calculations of positron scattering from molecular hydrogen. *J. Phys. Conf. Ser.* **2015**, *635*, 012009. [CrossRef]
179. Utamuratov, R.; Kadyrov, A.S.; Fursa, D.V.; Zammit, M.C.; Bray, I. Two-center close-coupling calculations of positron–molecular-hydrogen scattering. *Phys. Rev. A* **2015**, *92*, 032707. [CrossRef]
180. Weissbluth, M. *Atoms and Molecules*; Academic Press, Inc.: New York, NY, USA, 1978.
181. Pradhan, A.K.; Nahar, S.N. *Atomic Astrophysics and Spectroscopy*; Cambridge University Press: Cambridge, UK, 2011.
182. Joachain, C.J. *Quantum Collision Theory*; North-Holland: Amsterdam, Netherlands, 1983; Chapter 18.

183. Clementi, E.; Roetti, C. Roothaan-Hartree-Fock atomic wavefunctions: Basis functions and their coefficients for ground and certain excited states of neutral and ionized atoms, Z $\leq$ 54. *Atomic Data Nucl. Data Tables* **1974**, *14*, 177–478. [CrossRef]
184. O'Connel, J.K.; Lane, N.F. Non-adjustable exchange-correlation model for electron scattering from closed-shell atoms and molecules. *Phys. Rev. A* **1983**, *27*, 1983. [CrossRef]
185. Jain, A. New parameter-free polarization potentials in low wnergy positron collisions. In *Annihilation in Gases and Galaxies*; Drachman, R., Ed.; Goddard Space Flight Center: Greenbelt, MD, USA, 1989; p. 71.
186. Riley, M.E.; Truhlar, D.G. Approximations for the exchange potential in electron scattering. *J. Chem. Phys.* **1975**, *63*, 2182, doi:10.1063/1.431598. [CrossRef]
187. Chen, S.; McEachran, R.P.; Stauffer, A.D. Ab initio optical potentials for elastic electron and positron scattering from the heavy noble gases. *J. Phys. B* **2008**, *41*, 025201. [CrossRef]
188. Staszewsa, G.; Schwenke, D.W.; Thurumalai, D. Truhlar, D.G. Quasifree-scattering model for the imaginary part of the optical potential for electron scattering. *Phys. Rev. A* **1983**, *28*, 2740, doi:10.1103/PhysRevA.28.2740. [CrossRef]
189. Staszewsa, G.; Schwenke, D.W.; Truhlar, D.G. Investigation of the shape of the imaginary part of the optical-model potential for electron scattering by rare gases. *Phys. Rev. A* **1984**, *29*, 3078, doi:10.1103/PhysRevA.29.3078. [CrossRef]
190. Fuss, M.C.; Sanz, A.G.; Blanco, F.; Vieira, P.L.; Brunger, M.J.; Garcia, G. Differential and integral electron scattering cross sections from tetrahydrofuran (THF) over a wide energy range: 1–10,000 eV. *Eur. Phys. J. D* **2014**, *68*, 161. [CrossRef]
191. Watson, G.N. *Theory of Bessel Functions*; Cambridge University Press: London, UK, 1958; p. 368.
192. Gradshteyn, I.S.; Ryzhik, I.M. *Tables of Series, Integrals and Products*; Academic Press: New York, NY, USA, 1965.
193. Gupta, D.; Antony, B. Electron impact ionization of cycloalkanes, aldehydes, and ketones. *J. Chem. Phys.* **2014**, *141*, 054303. [CrossRef]
194. Singh, S.; Gupta, D.; Antony, B. Plasma relevant electron scattering cross sections of propene. *Plasma Sources Sci. Technol.* **2018**, *27*, 105014. [CrossRef]
195. Singh, S.; Naghma, R.; Kaur, J.; Antony, B. Calculation of total and ionization cross sections for electron scattering by primary benzene compounds. *J. Chem. Phys.* **2016**, *145*, 034309. [CrossRef]
196. Singh, S.; Antony, B. Study of inelastic channels by positron impact on simple molecules. *J. Appl. Phys.* **2017**, *121*, 244903. [CrossRef]
197. Reid, D.D.; Wadehra, J.N. A quasifree model for the absorption effects in positron scattering by atoms. *J. Phys. B* **1996**, *29*, L127 [CrossRef]
198. Singh, S.; Antony, B. Positronium formation and ionization of atoms and diatomic molecules by positron impact. *Europhys. Lett.* **2017**, *119*, 50006 [CrossRef]
199. Wigner, E.P. Resonance Reactions. *Phys. Rev.* **1946**, *70*, 606–618. [CrossRef]
200. Wigner, E.P.; Eisenbud, L. Higher Angular Momenta and Long Range Interaction in Resonance Reactions. *Phys. Rev.* **1947**, *72*, 29–40. [CrossRef]
201. Burke, P.G.; Hibbert, A.; Robb, W.D. Electron scattering by complex atoms. *J. Phys. B Atomic Mol. Phys.* **1971**, *4*, 153–161. [CrossRef]
202. Burke, P.G.; Robb, W.D. The R-Matrix Theory of Atomic Processes. *Adv. Atomic Mol. Phys.* **1975**, *11*, 143–214.
203. Faure, A. Gorfinkiel, J.D.; Morgan, L.A.; Tennyson, J. GTOBAS: Fitting continuum functions with Gaussian-type orbitals. *Comput. Phys. Commun.* **2002**, *144*, 224–241. [CrossRef]
204. Bloch, C. Une formulation unifiée de la théorie des réactions nucléaires. *Nucl. Phys.* **1957**, *4*, 503–528. [CrossRef]
205. Gailitis, M. New forms of asymptotic expansions for wavefunctions of charged-particle scattering. *J. Phys. B Atomic Mol. Phys.* **1976**, *9*, 843–854. [CrossRef]
206. Morgan, L.A. A generalized R-matrix propagation program for solving coupled second-order differential equations. *Comput. Phys. Commun.* **1984**, *31*, 419–422. [CrossRef]
207. Noble, C.J.; Nesbet, R.K. CFASYM, a program for the calculation of the asymptotic solutions of the coupled equations of electron collision theory. *Comput. Phys. Commun.* **1984**, *33*, 399–411. [CrossRef]
208. Burke, P.G.; Joachain, C.J. *Theory of Electron–Atom Collisions*; Springer: New York, NY, USA, 1995.

209. Sanna, N.; Gianturco, F.A. Differential cross sections for electron/positron scattering from polyatomic molecules. *Comp. Phys. Commun.* **1998**, *114*, 142–167. [CrossRef]
210. Brunger, M.J.; Buckman, S.J.; Ratnavelu, K. Positron Scattering from Molecules: An Experimental Cross Section Compilation for Positron Transport Studies and Benchmarking Theory. *J. Phys. Chem. Ref. Data* **2017**, *46*, 023102. [CrossRef]
211. Ratnavelu, K.; Brunger, M.J.; Buckman, S.J. Recommended Positron Scattering Cross Sections for Atomic Systems. *J. Phys. Chem. Ref. Data* **2019**, *48*, 023102. [CrossRef]
212. Srivastava, S.K.; Tanaka, H.; Chutjian, A.; Trajmar, S. Elastic scattering of intermediate-energy electrons by Ar and Kr. *Phys. Rev. A* **1981**, *23*, 2156. [CrossRef]
213. Sun, J.; Yu, G.; Jiang, Y.; Zhang, S. Total cross-sections for positron scattering by a series of molecules. *Eur. Phys. J. D* **1998**, *4*, 83–88. [CrossRef]
214. Hamada, A.; Sueoka, O. Total cross section measurements for positrons and electrons colliding with molecules. II. HCl. *J. Phys. B Atomic Mol. Opt. Phys.* **1994**, *27*, 5055-5064. [CrossRef]
215. Masin, Z.; Gorfinkiel, J.D.; Jones, D.B.; Bellm, S.M.; Brunger, M.J. Elastic and inelastic cross sections for low-energy electron collisions with pyrimidine. *J. Chem. Phys.* **2012**, *136*, 144310. [CrossRef]
216. Ferraz, J.; dos Santos, A.; de Souza, G.; Zanelato, A.; Alves, T.; Lee, M.-T.; Brescansin, L.; Lucchese, R.; Machado, L. Cross sections for electron scattering by formaldehyde and pyrimidine in the low-and intermediate energy ranges. *Phys. Rev. A* **2013**, *87*, 032717. [CrossRef]
217. Baek, W.Y.; Arndt, A.; Bug, M.; Rabus, H.; Wang, M. Total electronscattering cross sections of pyrimidine. *Phys. Rev. A* **2013**, *88*, 032702. [CrossRef]
218. Palihawadana, P.; Boadle, R.; Chiari, L.; Erson, E.; Machacek, J.; Brunger, M.J.; Buckman, S.; Sullivan, J. Positron scattering from pyrimidine. *Phys. Rev. A* **2013**, *88*, 012717. [CrossRef]
219. Bug, M.U.; Baek, W.Y.; Rabus, H.; Villagrasa, C.; Meylan, S.; Rosenfeld, A.B. An electron-impact cross section data set (10 eV–1 keV) of DNA constituents based on consistent experimental data: A requisite for Monte Carlo simulations. *Radiat. Phys. Chem.* **2017**, *130*, 459–479. [CrossRef]
220. Fuss, M.C.; Sanz, A.G.; Blanco, F.; Oller, J.C.; Limao-Vieira, P.; Brunger, M.J.; Garcia, G. Total electron-scattering cross sections from pyrimidine as measured using a magnetically confined experimental system. *Phys. Rev. A* **2013**, *88*, 042702. [CrossRef]
221. Sullivan, J.P.; Makochekanwa, C.; Jones, A.; Caradonna, P.; Slaughter, D.S.; Machacek, J.; McEachran, R.P.; Mueller, D.W.; Buckman, S.J. Forward angle scattering effects in the measurement of total cross sections for positron scattering. *J. Phys. B Atomic Mol. Opt. Phys.* **2011**, *44*, 035201. [CrossRef]
222. Sinha, N.; Antony, B. Electron and positron interaction with pyrimidine: A theoretical investigation. *J. Appl. Phys.* **2018**, *123*, 124906. [CrossRef]
223. Singh, S.; Gupta, D.; Antony, B. Electron and positron scattering cross sections for propene. *J. Appl. Phys.* **2018**, *124*, 034901. [CrossRef]
224. Singh, S.; Sen, A.; Antony, B. Positron scattering calculations of elastic, total and momentum transfer cross section for alkaline earth atoms. *Int. J. Mass Spectrom.* **2018**, *428*, 22–28. [CrossRef]
225. Singh, S.; Dutta, S.; Naghma, R.; Antony, B. Positron scattering from simple molecules. *J. Phys. B Atomic Mol. Opt. Phys.* **2017**, *50*, 135202. [CrossRef]
226. Sinha, N.; Singh, S.; Antony, B. Theoretical study of positron scattering by group 14 tetra hydrides: A quantum mechanical approach. *Int. J. Quant. Chem.* **2018**, *118*, e25679. [CrossRef]
227. Sinha, N.; Modak, P.; Singh, S.; Antony, B. Positron scattering from methyl halides. *J. Phys. Chem. A* **2018**, *122*, 2513–2522. [CrossRef]
228. Singh, S.; Antony, B. Positron induced scattering cross sections for hydrocarbons relevant to plasma. *Phys. Plasmas* **2018**, *25*, 053503. [CrossRef]
229. Sinha, N.; Antony, B. Inelastic cross sections for pentane isomers by positron impact. *Mol. Phys.* **2019**, *117*, 2527–2534. [CrossRef]
230. Sinha, N.; Antony, B. Theoretical study of positron scattering from pentane isomers. *Chem. Phys. Lett.* **2018**, *713*, 282–288. [CrossRef]

Article

Deep Minima in the Triply Differential Cross Section for Ionization of Atomic Hydrogen by Electron and Positron Impact

C. M. DeMars [1], S. J. Ward [1,*], J. Colgan [2], S. Amami [3] and D. H. Madison [3]

1 Department of Physics, University of North Texas, Denton, TX 76203, USA; CodyDeMars@my.unt.edu
2 Theoretical Division, Los Alamos National Laboratory, Los Alamos, NM 87545, USA; jcolgan@lanl.gov
3 Department of Physics, Missouri University of Science and Technology, Rolla, MI 65409, USA; samc5@mst.edu (S.A.); madison@mst.edu (D.H.M.)
* Correspondence: Sandra.Quintanilla@unt.edu

Received: 06 May 2020; Accepted: 25 May 2020; Published: 29 May 2020

Abstract: We investigate ionization of atomic hydrogen by electron- and positron-impact. We apply the Coulomb–Born (CB1) approximation, various modified CB1 approximations, the three body distorted wave (3DW) approximation, and the time-dependent close-coupling (TDCC) method to electron-impact ionization of hydrogen. For electron-impact ionization of hydrogen for an incident energy of approximately 76.45 eV, we obtain a deep minimum in the CB1 triply differential cross section (TDCS). However, the TDCC for 74.45 eV and the 3DW for 74.46 eV gave a dip in the TDCS. For positron-hydrogen ionization (breakup) we apply the CB1 approximation and a modified CB1 approximation. We obtain a deep minimum in the TDCS and a zero in the CB1 transition matrix element for an incident energy of 100 eV with a gun angle of 56.13°. Corresponding to a zero in the CB1 transition matrix element, there is a vortex in the velocity field associated with this element. For both electron- and positron-impact ionization of hydrogen the velocity field rotates in the same direction, which is anticlockwise. All calculations are performed for a doubly symmetric geometry; the electron-impact ionization is in-plane and the positron-impact ionization is out-of-plane.

Keywords: electron-impact ionization; hydrogen; positron-impact ionization; velocity field; vortices

---

1. Introduction

Studies of the angular distributions of ionized atomic electrons by charged-particle impact is a rich field that has long been studied due to its importance to other areas of physics (e.g., plasma, medical physics) and also due to the information that is available about the correlated nature of the particle interactions.

It was thought that structures in differential cross sections of atoms by the impact of fast bare charged particles could be described by using the first and second Born terms [1]. However, as discussed by Berakdar and Briggs [2] the deep minimum observed in the experimentally measured triply differential cross section (TDCS) of helium [3,4] is a different feature to those structures reported in reference [1]. Interestingly, Macek et al. [5] showed that a zero in the ionization element and the TDCS obtained using the theory of reference [2,6] corresponds to a vortex in the velocity field associated with this element.

Berakdar and Briggs [2] for electron-impact ionization of helium used the product of three Coulomb waves for the final state [6]. They expressed the transition matrix element T_{fi} as the sum of three amplitudes, T_1, T_2 and T_3, which represent, respectively, the initial scattering of the incoming electron with the active electron, the nucleus and the passive electron of the helium atom. In their paper [2], they explained that, in order for both the real and imaginary part of T_{fi} to be zero, the three amplitudes have to destructively interfere and that the amplitude T_3 is essential for a zero. Since for electron-hydrogen ionization the term $T_3 = 0$, as there is no passive electron in the target atom, Berakdar and Briggs [2] deduced that there cannot be an exact zero in T_{fi} and thus in the TDCS. Nevertheless, as pointed out by Berakdar and Briggs [2], Berakdar and Klar [7] obtained a dip in the TDCS for positron-hydrogen ionization. Also, recently, Navarrete et al. [8] and Navarrete and Barrachina [9–11], using for the final state the correlated three Coulomb wave (C3) wave function, have established that there are zeros in the transition matrix element for positron-hydrogen ionization, whereas, of course, there is also no passive electron. So, there exists the open question of whether it is possible to have zeros in the transition matrix element and the TDCS for hydrogen ionization by electron impact as well as by positron impact.

We apply the Coulomb–Born (CB1) [12–15] and modified CB1 [13,16] approximations to both positron- and electron-impact ionization of atomic hydrogen to see if we find a zero in the CB1 transition matrix $T^{\mathrm{CB1}}_{\mathbf{k},1s}$ [17–20]. We consider direct ionization (breakup) by positron impact only and, thus, we do not consider positronium formation into the continuum [21,22]. Electron-hydrogen ionization and positron-hydrogen ionization are fundamental three-body Coulomb processes. An exact zero in the ionization transition matrix element means that there is a deep minimum in the TDCS and a corresponding vortex in the velocity field, $\mathbf{v}$, associated with this element [5,18,19,23–27]. A significant advantage of the CB1 method over more sophisticated methods is that CB1 calculations can be performed rapidly, enabling a systematic search for zeros in T_{fi} to be done quickly and for the velocity field $\mathbf{v}$ associated with T_{fi} to be readily computed. Once a zero in $T^{\mathrm{CB1}}_{\mathbf{k},1s}$ and in the corresponding TDCS is located, more elaborate methods can be applied for the kinematics, or approximately the kinematics, of the zero in $T^{\mathrm{CB1}}_{\mathbf{k},1s}$. In addition to applying the CB1 and various modified CB1 approximations to electron-hydrogen ionization, we also apply the three body distorted wave (3DW) approximation and the time-dependent close-coupling (TDCC) method to electron-hydrogen ionization at incident energies close to where we obtained a zero in $T^{\mathrm{CB1}}_{\mathbf{k},1s}$.

The 3DW and TDCC methods have earlier been applied for electron-impact ionization of atomic hydrogen, molecular hydrogen and helium [28]. For the kinematics considered in reference [28], minima in the TDCSs for electron-impact ionization of atomic hydrogen were obtained, but these minima are not deep or zero. However, the TDCC [28] method obtained a deep minimum in the TDCS for electron-helium ionization for an incident energy of 64.6 eV that compared very well with experimental measurements [4]. (This incident energy of 64.6 eV is about three times the binding energy of helium.) The 3DW TDCS for the incident energy of 64.6 eV shows only a dip, although it does give a strong minimum for the incident energies of 44.6 and 54.6 eV [28]. Also, in the region of the experimental data [4], the CB1 and modified CB1 TDCSs for electron-helium ionization compared reasonably well with the measurements and with the TDCC results [25,28].

Previously, the CB1 approximation has been applied to electron-impact ionization of a K-shell model carbon atom at a high energy of 1801.2 eV [24]. A deep minimum in the CB1 TDCS and a zero in $T^{\mathrm{CB1}}_{\mathbf{k},1s}$ was obtained. Corresponding to the zero in $T^{\mathrm{CB1}}_{\mathbf{k},1s}$ there is vortex in the velocity field associated with this element [24].

For positron-hydrogen ionization a deep minimum in the fully differential cross section has previously been explained in terms of a vortex in the generalized velocity field $\mathbf{u}$ that is associated with the transition matrix element $T(\mathbf{k}_+, \mathbf{k}_-)$, where $T(\mathbf{k}_+, \mathbf{k}_-)$ depends on the momentum of the scattered positron $\mathbf{k}_+$ and the momentum of the ejected electron $\mathbf{k}_-$ [8,9]. Navarrete and Barrachina [9–11] obtained zeros in $T(\mathbf{k}_+, \mathbf{k}_-)$ and found that the velocity field rotates in opposite directions around the zeros. Recently, a deep minimum in the TDCS for positron-helium ionization was obtained using the CB1 and modified CB1 approximations [18–20,25–27].

Positron-impact ionization of atoms and molecules is of experimental interest [29–35], including studies of differential cross sections [36,37]. Structures in differential cross sections are of both theoretical and experimental interest [38].

In this paper we show that the CB1 and modified CB1 approximations give a deep minimum in the TDCS in the doubly symmetric geometry [4] for ionization of atomic hydrogen by both electron and positron impact. For electron impact the incident energy is about six times the binding energy, whereas for positron impact the incident energy is approximately seven times the binding energy. We determine the velocity field $\mathbf{v}$ associated with $T^{\rm CB1}_{\mathbf{k},1s}$ for both projectiles and notice that there is a vortex in this field which is seen by the swirling of the field around a zero in $T^{\rm CB1}_{\mathbf{k},1s}$. We are not able to compare our results with Navarrete and Barrachina's results [9–11] since they used collinear geometry and we used the doubly symmetric out-of-plane geometry.

We compare for electron-impact ionization of hydrogen the CB1, various modified CB1, 3DW and the TDCC TDCSs. This comparison is valuable since the CB1 approximation, a distorted-wave approximation, is perturbative and is applicable at high energies, while the TDCC method is an ab-initio method that is typically applied at low to intermediate energies. The CB1 gives the correct asymptotic limit for the ionization amplitude for fixed scattering angle. However, the CB1 TDCS multiplied by the modulus squared of the normalization factor of the Coulomb wave function of the two outgoing electrons gives the correct behavior in the vicinity of the Wannier's threshold law [13]. This approximation that obtains the TDCS as the product of the CB1 and the modulus squared of the normalization factor is called as the improved final-state Coulomb–Born approximation (ICBA) [13], and it is one of the three modified Coulomb–Born (CB1) approximations given in reference [16]. Surprisingly, the 3DW method, while a perturbative method, can give results that are good all the way down close to threshold, depending on the final-state energy. Since the 3DW has the post collision interaction (PCI) to all orders of perturbation theory, the 3DW can work well for intermediate to low energies, depending on the strength of the PCI [39,40].

The outline of the paper is as follows. In Section 2 we briefly describe the CB1, various modified CB1, the 3DW and the TDCC methods. We also give the expression for the velocity field $\mathbf{v}$ associated with $T^{\rm CB1}_{\mathbf{k},1s}$ and the equation of the circulation Γ. In Section 3.1 we compare for electron-hydrogen ionization the TDCS computed with the various methods. We also show $T^{\rm CB1}_{\mathbf{k},1s}$ and direction of the velocity field $\hat{\mathbf{v}} = \mathbf{v}/|\mathbf{v}|$. In Section 3.2 we present for positron-hydrogen ionization the TDCS computed with the CB1 and modified CB1 approximations and we show both $T^{\rm CB1}_{\mathbf{k},1s}$ and the direction of the velocity field.

We use atomic units throughout the paper unless otherwise stated. We report angles in degrees and the incident energies in eV.

2. Theory

The CB1 transition matrix element $T^{\rm CB1}_{\mathbf{k},1s}$, derived by Botero and Macek [12–15], is the first non-zero term in a perturbative expansion of the exact transition matrix element. For electron-impact ionization from atomic hydrogen, the direct $T^{\rm CB1}_{\mathbf{k},1s}$ is given by [12–15]

$$T^{\rm CB1}_{\mathbf{k},1s} = \langle \psi^{-}_{\mathbf{K}_f}(\mathbf{r})\psi^{-}_{\mathbf{k}}(\mathbf{r}')\left|\frac{1}{|\mathbf{r}-\mathbf{r}'|}\right|\varphi_i(\mathbf{r}')\psi^{+}_{\mathbf{K}_i}(\mathbf{r})\rangle, \tag{1}$$

where $\mathbf{k}$, $\mathbf{K}_i$ and $\mathbf{K}_f$ are the momentum of the ejected electron, the momentum of the incident particle, and momentum of the scattered particle, respectively. In Equation (1) $\mathbf{r}$ and $\mathbf{r}'$ are, respectively, the position vector of the incident (or scattered) particle and of the atomic (or ejected) electron relative to the proton. Also, in this equation, φ_i is the ground-state wave function of hydrogen and $\psi^{\pm}$ are Coulomb wave functions where the subscript $-$ and the subscript $+$ refer to incoming and outgoing boundary conditions, respectively [14,16].

For the doubly symmetric geometry [3,4,41] and for the effective charge $Z_{\rm eff}$ in the Coulomb wave functions of the incident electron and the scattered electron equal to the atomic number $Z_T = 1$ of hydrogen, the direct and exchange CB1 transition matrix elements are equal. Therefore, the CB1 TDCS for electron-hydrogen ionization can be expressed as:

$$\frac{d^3\sigma^{\rm CB1}}{d\Omega_f dE_{\mathbf{k}} d\Omega_{\mathbf{k}}} = (2\pi)^4 \frac{K_f k}{K_i} |T^{\rm CB1}_{\mathbf{k},1s}|^2, \tag{2}$$

where $d\Omega_{\mathbf{k}}$ is the solid angle for the ejected electron, $d\Omega_f$ is the solid angle for the scattered particle, and $E_{\mathbf{k}}$ is the energy of the ejected electron [14,24].

The modified CB1 TDCS for electron-impact ionization is the CB1 TDCS multiplied by the normalization factor given by reference [16], namely

$$|D_3^-(\mathbf{r}_{3ave})|^2 = \frac{\pi \exp(-\frac{\pi}{k_3})}{k_3[1-\exp(-\frac{\pi}{k_3})]} |{}_1F_1(\frac{i}{2k_3}, 1, -2ik_3 r_{3ave})|^2, \tag{3}$$

where $k_3 = |\mathbf{K}_f - \mathbf{k}|/2$. Also, $\mathbf{r}_{3ave}$ is the vector whose magnitude is the average spacing between the two outgoing electrons and whose direction is taken to be along $\hat{\mathbf{k}}_3$, $\mathbf{r}_{3ave} = r_{3ave}\hat{\mathbf{k}}_3$, which is appropriate in the Wannier region.

We follow reference [16] by considering three modified CB1 approximations which correspond to three different values of r_{3ave}, namely $r_{3ave} = 0$, $r_{3ave} = 1/k_3$ and

$$r_{3ave} = \frac{\pi^2}{16\varepsilon}\left(1 + \frac{0.627}{\pi\sqrt{\varepsilon} ln \varepsilon}\right)^2, \tag{4}$$

respectively. In Equation (4), $\varepsilon = (K_f^2 + k^2)/2$, which is the total energy of the two outgoing electrons. By taking the limit $r_{3ave} \to \infty$, $|D_3^-(\mathbf{r}_{3ave})|^2 \to 1$, the CB1 TDCS is obtained. In contrast, when $r_{3ave} = 0$, $|D_3^-(\mathbf{r}_{3ave})|^2 = |N^-_{e^-e^-}|^2$, where $N^-_{e^-e^-}$ is the normalization factor of the Coulomb wave function $\psi^-_{\mathbf{k}_3}$ of the two outgoing electrons. This modified Coulomb–Born approximation is called the ICBA in reference [13].

The modulus squared of the normalized factor of the Coulomb wave function for two outgoing particles (e^- and e^- for electron-impact ionization and e^+ and e^- for positron-impact ionization) is given by [7,13,16]

$$|N^-_{e^-e^\pm}|^2 = \frac{\pm\pi \exp(\pm\frac{\pi}{k_3})}{k_3[\exp(\pm\frac{\pi}{k_3}) - 1]}. \tag{5}$$

In an earlier paper [25], and for positron-impact ionization of hydrogen, we refer to the CB1 TDCS multiplied by $|N^-_{e^-e^\pm}|^2$ as the modified CB1 TDCS. Here, where we are considering three different modified CB1 TDCS for electron-impact ionization of hydrogen, we specify this particular modified CB1 TDCS as the modified CB1 TDCS with $r_{3ave} = 0$ to distinguish it from the other two (where $r_{3ave} = 1/k_3$ and r_{3ave} is given by Equation (4)).

For positron-impact ionization, $\mathbf{K}_i$ and $\mathbf{K}_f$ in Equations (1) and (2) are the momentum of the incident positron and the momentum of the scattered electron, respectively, and the interaction of the positron with the proton is repulsive.

Triple differential cross sections for electron-impact ionization of hydrogen are also calculated within the 3DW approximation. The 3DW approach is described in previous publications [42–44] (and references therein), so here we only present the theoretical aspects relevant to the present paper. In this approach, the wave functions for the incident electron, scattered electron, and ejected electron are all distorted waves. The important distinction of the method is that the exact final state electron-electron interaction, normally called the post collision interaction (PCI), is included in the approximation for the final state wave function of the system. The fact that any physical effect included in the system wave function is automatically included to all orders of perturbation theory means that PCI is included to all orders of perturbation theory in this approach. In the 3DW approximation the direct and exchange amplitudes are identical for the doubly symmetric in-plane geometry for electron-hydrogen ionization.

The TDCC calculations shown here are very similar to previous calculations [28] for electron ionization of atomic hydrogen. In brief, the TDCC approach solves the time-dependent Schrödinger equation for the two electrons by propagating the corresponding differential equations in time until convergence of the probabilities for ionization has been reached. The calculations shown here included fourteen partial waves, which were found to be sufficient for the TDCS calculations shown here. We note that the small cross sections found at larger scattering angles (beyond 100°) are quite sensitive to some of the numerical aspects of the calculations (such as mesh size, number of coupled channels). We note that these small cross sections are around three orders of magnitude lower than the peak of the cross section, which occurs at a scattering angle of approximately 39°.

When there is a zero in the real and imaginary parts of $T_{\mathbf{k},1s}$, there is a vortex in the velocity field associated with $T_{\mathbf{k},1s}$. Previously vortices in the velocity fields associated with atomic wave functions have been discussed by Bialynicki-Birula et al. [45]. Recently vortices in the velocity field associated with the ionization amplitude have been discussed by Macek et al. [5] and by Macek [23]. Here, we discuss vorticies in the velocity field associated with the transition matrix element which is proportional to the ionization amplitude [24]. The velocity field $\mathbf{v}$ associated with $T^{CB1}_{\mathbf{k},1s}$ for electron or positron ionization can be expressed as [24,25]

$$\mathbf{v} = \nabla_{\mathbf{k}} \mathrm{Im}[\ln T^{CB1}_{\mathbf{k},1s}]. \tag{6}$$

The circulation, Γ, for a closed contour, c, taken in the anticlockwise direction that encloses only one first-order zero in $T^{CB1}_{\mathbf{k},1s}$ is given by [5,23–25,45]

$$\Gamma = \int_c \mathbf{v} \cdot d\ell = \pm 2\pi. \tag{7}$$

3. Results

3.1. Deep Minimum in the TDCS for Electron-Impact Ionization of Hydrogen

For electron-impact ionization of hydrogen we perform calculations for the doubly symmetric in-plane geometry (Figure 1a) where both outgoing particles have the same energy and same angle, ξ, with respect to the z-axis. This geometry was used by Murray and Reed in their experimental measurements for electron-helium ionization [4]. The scattered electron makes a polar angle with respect to the z-axis of ξ while the ejected electron makes an angle of $-\xi$ with respect to the same axis.

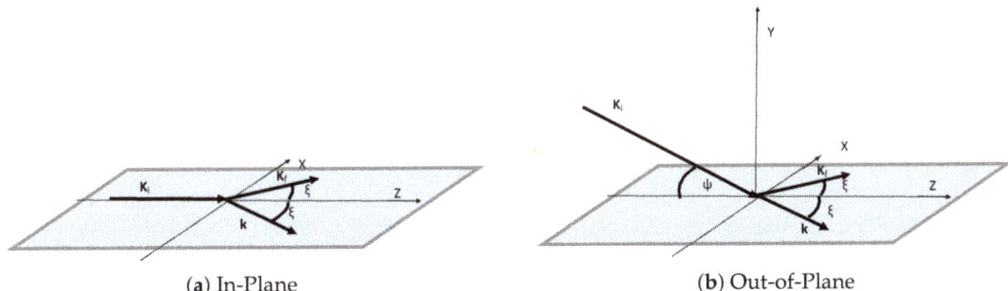

(a) In-Plane (b) Out-of-Plane

Figure 1. Geometry figures for both in-plane and out-of-plane configurations.

A deep minimum in the TDCS occurs when the real and imaginary parts of the transition matrix element, $T_{\mathbf{k},1s}$, are identically zero at the same angle; this occurs at a polar angle of 87.94° in the CB1 approximation (Figure 2). A comparison of the TDCSs computed with the CB1, modified CB1 with $r_{3ave} = 0$, modified CB1 with $r_{3ave} = 1/k_3$, and the modified CB1 with r_{3ave} given by Equation (4) approximations can be seen in Figure 3a. The position of the minimum is the same for these four approximations.

We show the comparison between the modified CB1 with $r_{3ave} = 0$, 3DW, and TDCC methods in Figure 3b. There is a dip 3DW TDCS at a polar angle of 90.2° and a slight dip in the TDCC TDCS at a polar angle of 90.6°. The polar angle for the deep minimum in the CB1 TDCS is within 3 degrees of these angles.

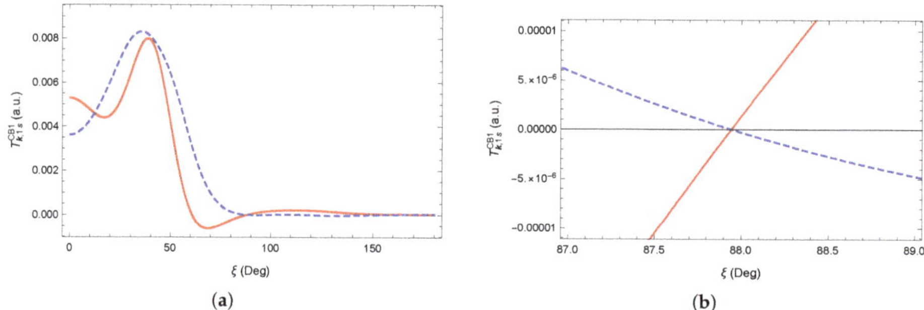

Figure 2. Transition matrix element, $T^{\rm CB1}_{\mathbf{k},1s}$, versus the polar angle ξ for electron-impact ionization of hydrogen at 76.4541 eV. (**a**) Re[$T^{\rm CB1}_{\mathbf{k},1s}$] (solid red line) and Im[$T^{\rm CB1}_{\mathbf{k},1s}$] (dashed blue line) over the full angular range (**b**) Re[$T^{\rm CB1}_{\mathbf{k},1s}$] (solid red line) and Im[$T^{\rm CB1}_{\mathbf{k},1s}$] (dashed blue line) in the vicinity of the zero in $T^{\rm CB1}_{\mathbf{k},1s}$.

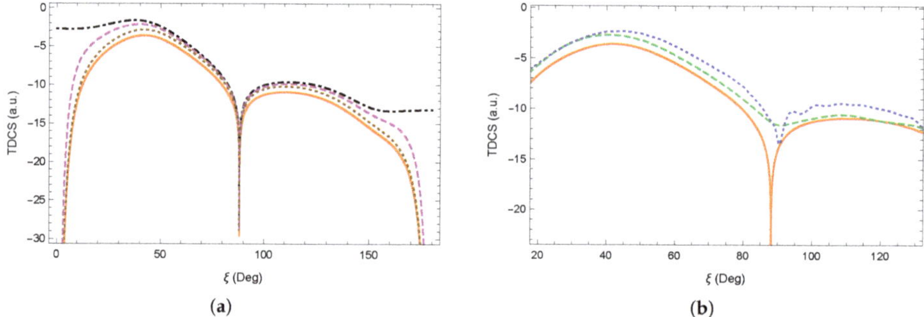

Figure 3. Triply differential cross section (TDCS) on a log (ln) scale versus the polar angle ξ for electron-impact ionization of hydrogen at 76.4541 eV. (**a**) Comparison of the CB1 TDCS (dot-dashed black line), the modified CB1 TDCS with $r_{3ave} = 0$ (solid orange line), the modified CB1 TDCS with $r_{3ave} = 1/k_3$ (dashed purple line) and the modified CB1 TDCS with r_{3ave} given by Equation (4) (dotted brown line). (**b**) Comparison of the modified CB1 TDCS with $r_{3ave} = 0$ (solid orange line) with an incident energy of 76.4541 eV, the 3DW TDCS (dotted blue line) for an incident energy of 76.46 eV, and the TDCC TDCS (dashed green line) for an incident energy of 76.45 eV.

A zero in the transition matrix element means that there is a vortex in the velocity field (Figure 4) associated with this element. The rotation of the velocity field is in the anticlockwise direction and the value of the circulation Γ is 2π. Figure 4 shows the nodal lines of Re[$T^{\rm CB1}_{\mathbf{k},1s}$] and Im[$T^{\rm CB1}_{\mathbf{k},1s}$] and it shows the direction of the velocity field, $\hat{\mathbf{v}} = \mathbf{v}/|\mathbf{v}|$, by the arrows. (The axes for this figure are k_z and k_x, where k_z and k_x are respectively the z- and x-components of the momentum $\mathbf{k}$ of the ejected electron [25].) The figure also shows a density plot of $\ln|T^{\rm CB1}_{\mathbf{k},1s}|$.

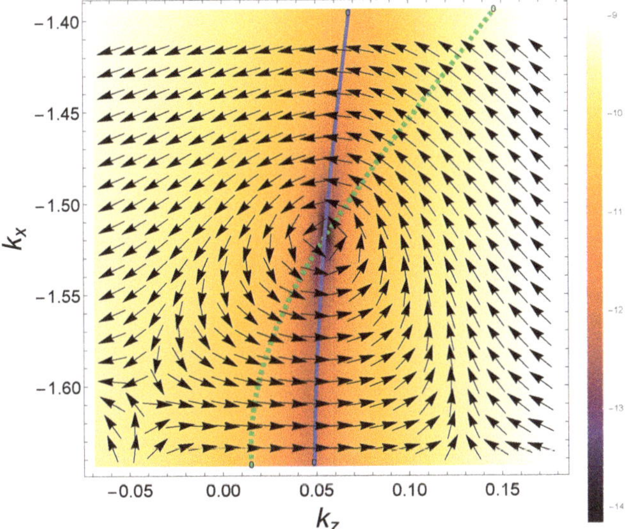

Figure 4. A density plot of $\ln|T^{\mathrm{CB1}}_{\mathbf{k},1s}|$ for electron-impact ionization of hydrogen for a fixed incident energy of 76.4541 eV in-plane and a grid in the z- and x-components of the momentum (k_z, k_x) of the ejected electron $\mathbf{k}$. The nodal lines of $\mathrm{Re}[T^{\mathrm{CB1}}_{\mathbf{k},1s}]$ and $\mathrm{Im}[T^{\mathrm{CB1}}_{\mathbf{k},1s}]$ are shown respectively, by the solid blue line and the dashed green line. The direction of the velocity field, $\hat{\mathbf{v}} = \mathbf{v}/|\mathbf{v}|$, is indicated by the arrows.

3.2. Deep Minimum in the CB1 TDCS for Positron-Impact Ionization of Hydrogen

For positron-impact ionization of hydrogen, we also consider a doubly symmetric geometry; however, for this projectile, we use the out-of-plane geometry that we show in Figure 1b. The gun angle ψ is the angle the incident particle makes with the z-axis [4]. We find a deep minimum in the CB1 TDCS for an incident energy of 100 eV with a gun angle of 56.13°. We choose this incident energy because it had been previously used [9]. Using the CB1 approximation, we vary the gun angle until we obtain a zero in $T^{\mathrm{CB1}}_{\mathbf{k},1s}$. The zero in $T^{\mathrm{CB1}}_{\mathbf{k},1s}$ can be seen in Figure 5.

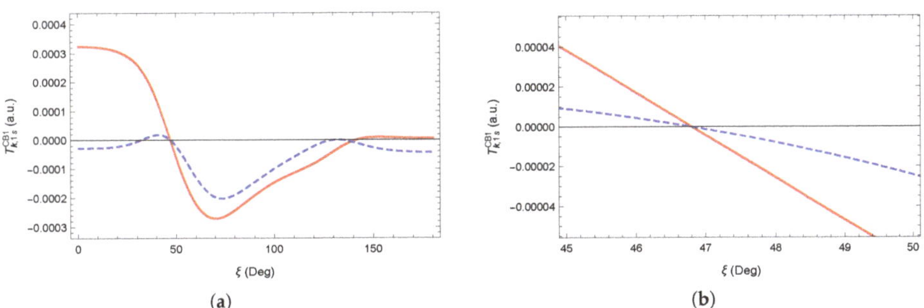

Figure 5. Transition matrix element $T^{\mathrm{CB1}}_{\mathbf{k},1s}$ versus the polar angle, ξ, for positron-impact ionization of hydrogen at 100 eV with a gun angle, ψ, of 56.13°. (**a**) $\mathrm{Re}[T^{\mathrm{CB1}}_{\mathbf{k},1s}]$ (solid red line) and $\mathrm{Im}[T^{\mathrm{CB1}}_{\mathbf{k},1s}]$ (dashed blue line) over the full angular range (**b**) $\mathrm{Re}[T^{\mathrm{CB1}}_{\mathbf{k},1s}]$ (solid red line) and $\mathrm{Im}[T^{\mathrm{CB1}}_{\mathbf{k},1s}]$ (dashed blue line) in the vicinity of the zero in $T^{\mathrm{CB1}}_{\mathbf{k},1s}$.

In the positron-impact ionization of hydrogen at 100 eV with a gun angle of 56.13° there is a deep minimum in the TDCS at a polar angle of 46.79°. This deep minima in the TDCS and a secondary dip around $\xi \approx 140°$ can be seen in Figure 6. The secondary dip in the TDCS is in the vicinity of the intersection point of the real and imaginary parts of $T^{\mathrm{CB1}}_{\mathbf{k},1s}$ and a second zero in $\mathrm{Re}[T^{\mathrm{CB1}}_{\mathbf{k},1s}]$ (see Figure 5). We perform an out-of-plane search to see if we could find a zero in $T^{\mathrm{CB1}}_{\mathbf{k},1s}$ around $\approx 140°$, but without success.

We note that we also obtain a deep minimum in the TDCS for a slightly lower incident energy, 95 eV, if the gun angle is reduced a little to 55.75°. We expect that we could track the position of the deep minimum for different incident energies and obtain a locus of points that gives the angles ψ and ξ for these energies as we did previously for electron-helium ionization [25].

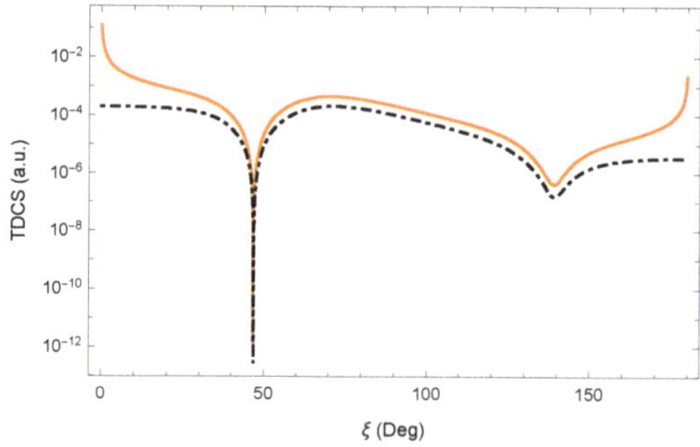

Figure 6. TDCS on a log (ln) scale versus the polar angle, ξ, for positron-impact ionization of hydrogen at 100 eV with a gun angle of 56.13°. Comparison of the CB1 TDCS (dot-dashed black line) and the modified CB1 TDCS (solid orange line) results.

For positron-impact ionization of hydrogen, we find that circulation Γ is 2π and the velocity field that is associated with the zero in $T^{\mathrm{CB1}}_{\mathbf{k},1s}$ rotates anticlockwise, as can be seen in Figure 7. This figure shows the direction of the velocity field, $\hat{\mathbf{v}} = \mathbf{v}/|\mathbf{v}|$, by the arrows and it shows the nodal lines of $\mathrm{Re}[T^{\mathrm{CB1}}_{\mathbf{k},1s}]$ and $\mathrm{Im}[T^{\mathrm{CB1}}_{\mathbf{k},1s}]$. It also gives a density plot of $\ln|T^{\mathrm{CB1}}_{\mathbf{k},1s}|$.

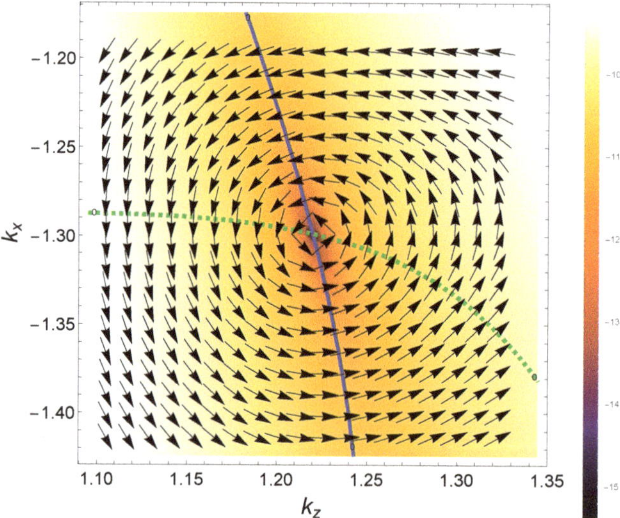

Figure 7. A density plot of $\ln|T^{\mathrm{CB1}}_{\mathbf{k},1s}|$ for positron-impact ionization of hydrogen for a fixed incident energy of 100 eV out-of-plane with a gun angle of 56.13° and a grid in the z- and x-components of the momentum (k_z, k_x) of the ejected electron $\mathbf{k}$. The nodal lines of $\mathrm{Re}[T^{\mathrm{CB1}}_{\mathbf{k},1s}]$ and $\mathrm{Im}[T^{\mathrm{CB1}}_{\mathbf{k},1s}]$ are shown respectively, by the solid blue line and the dashed green line. The direction of the velocity field, $\hat{\mathbf{v}} = \mathbf{v}/|\mathbf{v}|$, is indicated by the arrows.

For the specific geometries and kinematics that we consider, the direction of rotation of the velocity field is the same for positron- and electron-impact ionization of hydrogen. Interestingly, for the two ionization processes, the polar angle ξ of the deep minimum in the CB1 TDCS is less than $90°$, and thus, ξ lies in the first quadrant.

4. Conclusions

While a deep minimum in the TDCS was not expected to exist for electron-impact ionization of hydrogen, we were able to obtain one using the CB1 and modified CB1 approximations for an incident energy of 76.4541 eV [2]. The CB1 approximation is a high-energy approximation that has been applied to intermediate energies. Previously, the CB1 and modified CB1 approximations have been applied to electron-helium ionization for an incident energy of 64.6 eV [25]; and, in the region of the experimental data, the TDCS results compare reasonably well with experimental measurements [4] and with TDCC results [28]. These approximations have also been applied previously to positron-helium ionization [25].

Using the CB1 and a modified CB1 approximations we obtain a deep minimum in the TDCS for positron-impact ionization of hydrogen for an incident energy of 100 eV in the doubly symmetric out-of-plane geometry with a gun angle of $56.13°$. This energy has been previously considered for positron-hydrogen ionization in the collinear geometry [8,9]. A benefit of the CB1 approximation is that it is a fairly simple approximation that does not take much computing time.

While the question of whether there can be a zero in the transition matrix element and a corresponding deep minimum in the TDCS for electron-impact ionization of hydrogen has not been resolved, we have found that these exist within the CB1 approximation. Due to the fact that zeros in a transition matrix element have been found previously for positron-impact ionization of hydrogen [8,9], a zero in $T^{\mathrm{CB1}}_{\mathbf{k},1s}$ for electron-hydrogen ionization is not that surprising even though there is no passive electron. We note that, for electron-hydrogen ionization, neither the TDCC or 3DW methods obtained a deep minimum in the TDCS; however, they did obtain a dip.

The direction of the rotation of the velocity field around a zero in $T^{\mathrm{CB1}}_{\mathbf{k},1s}$ does not appear to be projectile dependent as for both electron- and positron-impact ionization of hydrogen the rotation is anticlockwise. This may be due to the fact that the polar angle of the zero in $T^{\mathrm{CB1}}_{\mathbf{k},1s}$ is between $0°$ and $90°$ (first quadrant) for both cases. An earlier paper shows, for positron-helium ionization, clockwise rotation of the velocity field about the zero in $T^{\mathrm{CB1}}_{\mathbf{k},1s}$; however, the polar angle for this zero is between $90°$ and $180°$ (second quadrant) [25]. For the energies that were considered in reference [25] for electron-helium ionization the direction of the velocity field is anticlockwise and the polar angle lies in the first quadrant.

The interesting finding that for both electron- and positron-impact ionization of hydrogen the velocity field rotates in the same direction is different from the previous finding for impact ionization of helium [25]. This may be due to the fact that for impact ionization of hydrogen by the two projectiles the deep minimum in the TDCS occurs at an angle less than $90°$, while for earlier results of electron and positron ionization of helium the zeros occur in different quadrants [25].

Author Contributions: C.M.D. and S.J.W. prepared the manuscript (with input from J.C. and D.H.M.). C.M.D., under the supervision of S.J.W., performed the CB1 and modified CB1 calculations for both electron- and positron-impact ionization of hydrogen. J.C. performed the TDCC calculations and provided a discussion of the method. The 3DW calculations are from D.H.M. and S.A. A discussion on the 3DW approximation was provided by D.H.M. All authors have read and agreed to the published version of the manuscript.

Funding: S.J.W. is thankful for support from the NSF under Grant No. PHYS-1707792.

Acknowledgments: J. B. Kent, while an undergraduate student at UNT and under the supervision of S. J. Ward, performed CB1 and modified CB1 calculations for electron-impact ionization of hydrogen. We acknowledge that, for electron-hydrogen ionization, he was the first to locate the position of the zero in the CB1 transition matrix element and the corresponding deep minimum in the TDCS. SJW would like to thank Gaetana (Nella) Laricchia for encouraging the CB1 investigation. CMD and SJW are thankful for the CB1 Fortran codes from Javier Botero. Wolfram Research [46] and Microsoft Publisher [47] were used to generate the images in this paper. Also, Wolfram Research [46] was used in some of the calculations. JC acknowledges the support by the US Department of Energy through the ASC PEM Program of the Los Alamos National Laboratory. Los Alamos National Laboratory is operated by Triad National Security, LLC, for the National Nuclear Security Administration of U.S. Department of Energy (Contract No. 89233218CNA000001).

Conflicts of Interest: The authors declare no conflict of interest.

Abbreviations

The following abbreviations are used in this manuscript:

MDPI	Multidisciplinary Digital Publishing Institute
DOAJ	Directory of open access journals
TDCS	Triply differential cross section
CB1	Coulomb–Born
TDCC	Time-dependent close-coupling
3DW	Three body distorted wave

References

1. Briggs, J.S. Cusps, dips and peaks in differential cross-sections for fast three-body Coulomb collisions. *Comments At. Mol. Phys.* **1989**, *23*, 155–174.
2. Berakdar, J.; Briggs, J.S. Interference effects in (e,2e)-differential cross sections in doubly symmetric geometry. *J. Phys. B At. Mol. Opt. Phys.* **1994**, *27*, 4271–4280. [CrossRef]
3. Murray, A.J.; Read, F.H. Exploring the helium (e,2e) differential cross section at 64.6 eV with symmetric scattering angles but non-symmetric energies. *J. Phys. B At. Mol. Opt. Phys.* **1993**, *26*, L359–L365. [CrossRef]
4. Murray, A.J.; Read, F.H. Evolution from the coplanar to the perpendicular plane geometry of helium (e,2e) differential cross sections symmetric in scattering angle and energy. *Phys. Rev. A* **1993**, *47*, 3724–3732. [CrossRef] [PubMed]
5. Macek, J.H.; Sternberg, J.B.; Ovchinnikov, S.Y.; Briggs, J.S. Theory of deep minima in (e,2e) measurements of triply differential cross sections. *Phys. Rev. Lett.* **2010**, *104*, 033201. [CrossRef]
6. Berakdar, J.; Briggs, J.S. Three-body Coulomb continuum problem. *Phys. Rev. Lett.* **1994**, *72*, 3799–3802. [CrossRef] [PubMed]
7. Berakdar, J.; Klar, H. Structures in triply and doubly differential ionization cross sections of atomic hydrogen. *J. Phys. B At. Mol. Opt. Phys.* **1993**, *26*, 3891–3913 [CrossRef]
8. Navarrete, F.; Picca, R.D.; Fiol, J.; Barrachina, R.O. Vortices in ionization collisions by positron impact. *J. Phys. B At. Mol. Opt. Phys.* **2013**, *46*, 115203. [CrossRef]
9. Navarrete, F.; Barrachina, R.O. Vortices in the three-body electron-positron-proton continuum system induced by the positron-impact ionization of hydrogen. *J. Phys. B At. Mol. Opt. Phys.* **2015**, *48*, 055201. [CrossRef]
10. Navarrete, F.; Barrachina, R.O. Vortices in ionization collisions. *Nucl. Instrum. in Phys. Res. B* **2016**, *369*, 72–76. [CrossRef]
11. Navarrete, F.; Barrachina, R.O. Vortex rings in the ionization of atoms by positron impact. *J. Phys. B Conf. Ser.* **2017**, *875*, 012022. [CrossRef]
12. Botero, J.; Macek, J.H. Coulomb Born approximation for electron scattering from neutral atoms: Application to electron impact ionization of helium in coplanar symmetric geometry. *J. Phys. B At. Mol. Opt. Phys.* **1991**, *24*, L405–L411. [CrossRef]
13. Botero, J.; Macek, J.H. Threshold angular distributions of (e, 2e) cross sections of helium atoms. *Phys. Rev. Lett.* **1992**, *68*, 576–579. [CrossRef] [PubMed]
14. Botero, J.; Macek, J.H. Coulomb-Born calculation of the triple-differential cross section for inner-shell electron-impact ionization of carbon. *Phys. Rev. A* **1992**, *45*, 154–165. [CrossRef] [PubMed]
15. Macek, J.H.; Botero, J. Perturbation theory with arbitrary boundary conditions for charged-particle scattering: Application to (e,2e) experiments in helium. *Phys. Rev. A* **1992**, *45*, R8. [CrossRef]
16. Ward, S.J.; Macek, J.H. Wave functions for continuum states of charged fragments. *Phys. Rev. A* **1994**, *49*, 1049–1056. [CrossRef]

17. Ward, S.J.; Kent, J.B. Deep minimum in the Coulomb-Born TDCS for electron-impact ionization of atomic hydrogen. *Bull. Am. Phys. Soc.* **2017**, *62*, 28. Available online: http://meetings.aps.org/link/BAPS.2017.GEC.GT1.1 (accessed on 25 May 2020).
18. Ward, S.J. Vortices for Positron Ionization and Positronium Formation. Available online: http://meetings.aps.org/Meeting/GEC19/Session/LW1.6 (accessed on 25 May 2020).
19. DeMars, C.M.; Kent, J.B.; Ward, S.J. Abstract: Q01.00026: Deep Minima in the Coulomb-Born Triply Differential Cross Section for Electron and Positron Ionization of Hydrogen and Helium. Available online: https://meetings.aps.org/Meeting/DAMOP20/Session/Q01.26 (accessed on 25 May 2020).
20. DeMars, C.M.; Ward, S.J.; Kent, J.B. Deep Minima in the Coulomb-Born triply differential cross section for electron and positron ionization of hydrogen and helium. Submitted May 2020 to GEC 2020.
21. Kadyrov, A.S.; Bray, I.; Stelbovics, A.T. Near-threshold positron-impact ionization of atomic hydrogen. *Phys. Rev. Lett.* **2007**, *98*, 263202. [CrossRef]
22. Kadyrov, A.S.; Bray, I. Recent progress in the description of positron scattering from atoms using the convergent close-coupling theory. *J. Phys. B At. Mol. Opt. Phys.* **2016**, *49*, 22202. [CrossRef]
23. Macek, J.H. Vortices in atomic processes. In *Dynamical Processes in Atomic and Molecular Physics*; Ogurtsov, G., Dowek, D., Eds.; Bentham Science: Sharjah, UAE, 2012; Chapter 1, pp. 3–28.
24. Ward, S.J.; Macek, J.H. Effect of a vortex in the triply differential cross section for electron impact K-shell ionization of carbon. *Phys. Rev. A* **2014**, *90*, 062709. [CrossRef]
25. DeMars, C.M.; Kent, J.B.; Ward, S.J. Deep minima in the Coulomb-Born triply differential cross sections for ionization of helium by electron and positron impact. *Eur. Phys. J. D* **2020**, *74*, 48.
26. DeMars, C.M.; Kent, J.B.; Ward, S.J. Deep minimum in the Coulomb-Born TDCS for e^--He and e^+-He ionization. *Bull. Am. Phys. Soc.* **2019**, *64*, 119. Available online: http://meetings.aps.org/Meeting/DAMOP19/Session/L01.13 (accessed on 25 May 2020).
27. DeMars, C.M.; Ward, S.J. Deep minima in the TDCS for positron-helium ionization computed using the Coulomb-Born approximation. In *Proceedings of the XX International Workshop on Low-Energy Positron and Positronium Physics* (POSMOL 2019 Book of Abstracts), Belgrade, Serbia, 18–21 July 2019; Cassidy, D., Brunger, M.J., Petrovíc, Z.L., Dujko, S., Marinković, B.P., Maríc, D., Tošić, S., Eds.; Serbian Academy of Sciences and Arts: Belgrade, Serbia, 2019; LEPPP 12, p. 53. Available online: http://posmol2019.ipb.ac.rs/_files/Book_POSMOL2019_Online.pdf (accessed on 25 May 2020).
28. Colgan, J.; Al-Hagan, O.; Madison, D.H.; Murray, A.J.; Pindzola, M.S. Deep interference minima in non-coplanar triple differential cross sections for the electron-impact ionization of small atoms and molecules. *J. Phys. B At. Mol. Opt. Phys.* **2009**, *42*, 171001.
29. Jones, G.O.; Charlton, M.; Slevin, J.; Laricchia, G.; Kövér, Á.; Poulsen, M.R.; Chormaic, S.N. Positron impact ionization of atomic hydrogen. *J. Phys. B At. Mol. Opt. Phys.* **1993**, *26*, L483–L488. [CrossRef]
30. Murtagh, D.J.; Szłuińska, M.; Moxom, J.; Reeth, P.V.; Laricchia, G. Positron-impact ionization and positronium formation from helium. *J. Phys. B At. Mol. Opt. Phys.* **2005**, *38*, 3857–3866. [CrossRef]
31. Laricchia, G.; Brawley, S.; Cooke, D.A.; Kövér, Á.; Murtagh, D.J.; Williams, A.I. Ionization in positron- and positronium-collisions with atoms and molecules. *J. Phys. Conf. Ser.* **2009**, *194*, 012036.
32. Laricchia, G.; Cooke, D.A.; Köver, A.; Brawley, S.J. Experimental aspects of ionization studies by positron and positronium impact, In *Fragmentation Processes: Topics in Atomic and Molecular Physics*; Whelan, C.T., Ed.; Cambridge University Press: Cambridge, UK, 2013; Chapter 5, pp. 116–136.
33. DuBois, R.D. Topical Review, Methods and progress in studying inelastic interactions between positrons and atoms. *J. Phys. B At. Mol. Opt. Phys.* **2016**, *49*, 112002. [CrossRef]
34. Surko, C.M.; Gribakin, G.F.; Buckman, S.J. Topical Review, Low-energy positron interactions with atoms and molecules. *J. Phys. B At. Mol. Opt. Phys.* **2005**, *38*, R57–R126. [CrossRef]
35. Schippers, S.; Sokell, E.; Aumayr, F.; Sadeghpour, H.; Ueda, K.; Bray, I.; Bartschat, K.; Murray, A.; Tennyson, J.; Dorn, A.; et al. Roadmap on photonic, electronic and atomic collisions: II. Electron and antimatter interactions. *J. Phys. B At. Mol. Opt. Phys.* **2019**, *52*, 171002. [CrossRef]
36. Kövér, Á.; Laricchia, G. Triply differential study of positron impact ionization of H_2. *Phys. Rev. Lett.* **1998**, *80*, 5309–5312.
37. Kövér, Á.; Murtagh, D.J.; Williams, A.I.; Laricchia, G. Differential ionization studies by positron impact. *J. Phys. Conf. Ser.* **2010**, *199*, 012020. [CrossRef]

38. Navarrete, F.; Feole, M.; Barrachina, R.O.; Kövér, Á. When vortices and cusps meet. *J. Phys. Conf. Ser.* **2015**, *583*, 01202. [CrossRef]
39. Ali, E.; Ren, X.; Dorn, A.; Ning, C.; Colgan, J.; Madison, D.H. Experimental and theoretical triple-differential cross sections for tetrahydrofuran ionized by low-energy 26-eV-electron impact. *Phys. Rev. A* **2016**, *93*, 062705. [CrossRef]
40. Ali, E.; Chakraborty, H.S.; Madison, D.H. Improved theoretical calculations for electron-impact ionization of DNA analogue molecules. *J. Chem. Phys.* **2020**, *152*, 124303. [CrossRef] [PubMed]
41. Murray, A.J.; Read, F.H. Novel exploration of the helium (e,2e) ionization processes. *Phys. Rev. Lett.* **1992**, *69*, 2912–2914. [CrossRef] [PubMed]
42. Jones, S.; Madison, D.H.; Franz, A.; Altick, P.L. Three-body distorted-wave Born approximation for electron-atom ionization. *Phys. Rev. A* **1993**, *48* R22–R25. [CrossRef]
43. Jones, S.; Madison, D.H. Ionization of hydrogen atoms by fast electrons. *Phys. Rev. A* **2000**, *62*, 042701.
44. Madison, D.H.; Al-Hagan, O. The distorted-wave Born approach for calculating electron-impact ionization of molecules. *J. At. Mol. Opt. Phys.* **2010**, *2010*, 367180.
45. Bialynicki-Birula, I.; Bialynicka-Birula, Z.; Śliwa, C. Motion of vortex lines in quantum mechanics. *Phys. Rev. A* **2000**, *61*, 032110. [CrossRef]
46. Wolfram Research, Inc. *Mathematica*; Version 11.3; Wolfram Research, Inc.: Champaign, IL, USA, 2018.
47. Microsoft Office 365 ProPlus. Available online: https://www.microsoft.com/en-us/microsoft-365/publisher (accessed on 25 May 2020).

Article

Resonances in Systems Involving Positrons

Anand K. Bhatia

Heliophysics Science Division, NASA/Goddard Space Flight Center, Greenbelt, MD 20771, USA; anand.k.bhatia@nasa.gov

Received: 17 February 2020; Accepted: 5 May 2020; Published: 7 May 2020

Abstract: When an incident particle on a target gets attached to the target, the cross-section at that energy could be much larger compared to those at other energies. This is a short-lived state and decays by emitting an electron. Such states can also be formed by the absorption of a photon. Such states are below the higher thresholds and are called autoionization states, doubly excited states, or Feshbach resonances. There is also a possibility of such states to form above the thresholds. Then they are called shape resonances. Resonances are important in the diagnostic of solar and astrophysical plasmas. Some methods of calculating the resonance parameters are described and resonance parameters occurring in various systems are given.

Keywords: autoionization states; doubly excited states; Feshbach states; resonances; shape resonances

---

1. Introduction

In measuring scattering cross-sections, a peak or dip implies that the incident particle has formed a compound state which decays after a while. Such a state is called an autoionization state, a doubly excited state, or a resonance state, and has a very short lifetime compared to real bound states. Such states can also be formed by absorption of radiation in the target. Resonances are ubiquitous in electron-atom and electron-ion interactions. They play an important role in solar and astrophysical plasmas to infer temperatures and densities of plasmas [1]. However, they are not that common in the case of positron-target systems.

In a simple system like a positron-hydrogen, the positron and electron tend to be on the same side of the nucleus because of the attraction between the two particles, unlike in the case of the electron-hydrogen system where the two electrons tend to be on the opposite sides of the nucleus because of the repulsion between the two electrons. This shows that the correlations become very important in a positron-hydrogen system. This makes calculations of resonances difficult because many terms are required to calculate resonance parameters. Moreover, there is a positronium channel open below all the thresholds and this adds further complications.

2. Methods of Calculations

There are methods like the stabilization method, complex rotation method, Feshbach operator formalism [2], close-coupling method, and *R*-matrix method to calculate the resonance parameters. In approach [2], projection operators P and Q are defined such that P projects on a state and $Q = 1 - P$ removes that state, $P^2 = P$, $Q^2 = Q$ (idempotent), and $PQ = 0$ (orthogonality). We form a wave function $Q\Psi$ which is such that the lower states have been removed [3]. Therefore, using Raleigh–Ritz variational principle, we obtain eigenvalues:

$$\varepsilon_Q = \frac{< Q\Psi|H|Q\Psi >}{< Q\Psi|Q\Psi >} \quad (1)$$

These eigenvalues give us the positions of resonances. We obtain these bound states which are embedded in the continuum and are below the higher thresholds. They correspond to resonance states within the continuum which must be calculated separately and the width for each state must be calculated. In the narrow region of the width, the scattering phase shift increases by π radians. Calculation of the shift and width requires continuum functions (cf. Equation (2.13a)) in [3]. Various approximations like the exchange approximation, method of polarized orbitals, close-coupling approximation have been used to calculate continuum functions. However, it is difficult to write projection operators P and Q when the positronium channel is open. We need to use a method which does not depend on projection operator formalism [2].

The complex rotation method, based on a theorem by Belslev and Combes [4], has been applied extensively to calculate resonance parameters with great accuracy. The advantage of this formulation is that only discrete functions are included in the wave function and the continuum function is not necessary. In this method, the radial part is rotated by an angle θ. The Hamiltonian is transformed in the same way. The angle is varied until the eigenvalues do not change. Widths of the states are also obtained in the same calculation. Since there are also eigenvalues in the continuum, the shift mentioned above is included in the resonance positions obtained in this method. This is discussed further in Section 3.

This method gives the resonance positions which include the shift due to the interaction of discrete states with the continuum and need not be calculated separately. The eigenvalues obtained in this method are complex, where the complex part gives the width of the state.

In the positron-hydrogen system, Mittleman [5] showed that the equation for the positron-hydrogen system has an attractive potential proportional to $1/r^2$ due to the degeneracy of the 2s and 2p states of the hydrogen atom and therefore, there should be an infinite number of resonances in this case as in the electron-hydrogen system. Mittleman [5] conclusively showed the existence of resonances without carrying out detailed calculations. The resonances in electron-hydrogen system have been observed but not in the positron-hydrogen system, at least up to now.

3. Calculations and Results

The first successful calculation for the S-wave resonance was carried out by Doolen, Nuttal, and Wherry [6] using a sparse-matrix technique in the complex-rotation method. In this method, the radial coordinates are transformed by an angle θ:

$$r \to re^{i\theta} \tag{2}$$

The Hamiltonian $H = T + V$ is transformed to:

$$H = Te^{-2i\theta} + Ve^{-i\theta} \tag{3}$$

They used a wave function of the form:

$$\Phi(r_1, r_2, r_{12}) = \exp(-\alpha(r_1 + r_2)L_l^0(u)L_m^0(v)L_n^0(w) \tag{4}$$

In the above equation, α is the nonlinear parameter, L_l^0 is a Laguerre polynomial and:

$$u = \alpha(r_2 + r_{12} - r_1), v = \alpha(r_1 + r_{12} - r_2), \text{ and } w = 2\alpha(r_1 + r_2 - r_{12}) \tag{5}$$

They found only one resonance whose complex energy is given by:

$$E(complex) = \text{Re}(E) + \text{Im}(E) = \text{Re}(E) - i\Gamma/2 \tag{6}$$

Re(E) represents the position of the resonance and the imaginary part represents the half width of the resonance. In this method, Im(E) is plotted vs. Re(E) and we look for the stationary paths as the angle θ is increased for a fixed value of the nonlinear parameter α. Their results as a function of the number of terms are given in Table 1.

Table 1. Position and width of the resonance below $n = 2$ threshold of the hydrogen atom.

Number of Terms	Position	Γ/2
286	−0.2573744	0.0000676
364	−0.2573733	0.0000674
455	−0.2573745	0.0000671
560	−0.2573740	0.0000677
680	−0.2573741	0.0000677

This table shows that the resonance is at E = −0.2573741 Ry with a width of 0.0001354 Ry. It is clear from the table that a wave function with very large number of terms is needed to calculate the resonance parameters which is not so in the case of the electron-hydrogen. Prior to this, attempts by various authors failed to infer the existence of this resonance due to using a very small number of terms in their wave functions. This calculation provided incentive to look for resonances below the higher thresholds as well. Varga, Mitroy, Mezie, and Kruppa [7] carried out calculations below the n = 2, 3, and 4 thresholds. Their results are shown in Table 2. There are two resonances below $n = 2$ threshold, three below n = 3 threshold, and three below $n = 4$ threshold.

Table 2. Positron-hydrogen resonances [7] below higher thresholds. Units are Ry.

Threshold n	Position	Width
2	−0.257244	0.000132
	−0.250262	0.0000096
3	−0.116094	0.001284
	−0.112006	0.0003132
4	−0.076958	0.0000788
	−0.067714	0.0000528
	−0.064380	0.00003376

It is possible to find positions of higher resonances, using the relation between two resonances given by Temkin and Walker [8]:

$$\varepsilon_{n+1} = e^{-2\pi/\alpha}\varepsilon_n, \tag{7}$$

$$\alpha = \left(\sqrt{37} - 5/4\right)^{0.5} \tag{8}$$

This was deduced for electron-hydrogen resonances but does give reasonable values of higher resonances in positron-hydrogen system as well.

The potential between a positron and He^{+} is repulsive. Therefore, an existence of resonances in this system seems unlikely. Using the stabilization method, Bhatia and Drachman [9] showed the existence of several resonances. This was confirmed by Ho [10] who carried out a definitive calculation using the complex rotation method described above and showed the correctness of their results [9]. Hylleraas type functions have been used in most calculations. The resonance parameters obtained by him are shown in Table 3.

Table 3. Positions and widths of S- and P-resonances in e^+-He^+ system below the n = 2 threshold [10], Units Are Ry.

State	Position	Width
S-wave	−0.74099	0.25886
	−0.3712	0.0786
P-wave	−0.70869	0.35504
	−0.36956	0.08634

Kar and Ho [11] carried out similar calculations for e^+-He system and their results agree with those obtained by Ren, Han, and Shi [12], who used hyperspherical coordinates. Their results are shown in Table 4. The resonances are very narrow compared to those in e^+-He^+ system.

Table 4. Resonance parameters in the e^+-He system. Units Are Ry.

Resonance	Ren, Han, and Shi [12]		Kar and Ho [11]	
	Position	Width	Position	Width
1	−4.15308	0.00046	−4.15306	0.00052
2	−4.13262	0.00030	−4.1326	0.00034
4	−4.12556	0.00004		

4. Resonances in Ps

The Ps^- system is obtained when the proton in H^- is replaced by a positron. Now the nucleus has the same mass as an electron. Therefore, the mass polarization term in the Hamiltonian becomes important. The binding energy of Ps^- is very close to half of the binding energy of H^-. The Hamiltonian is given by:

$$H = -2\nabla_1^2 - 2\nabla_2^2 - 2\nabla_1 \cdot \nabla_2 - 2/r_1 - 2/r_2 + 2/r_{12} \quad (9)$$

$$H = T + V \quad (10)$$

where r_1, r_2, are the coordinates of electrons with respect to the positron and $r_{12} = \left|\vec{r}_1 - \vec{r}_{12}\right|$.

Resonances, in singlet and triplet states in Ps^-, have been calculated by Ho [13] using 364 terms in the singlet states and 455 terms in the triplet states. His results, obtained using the complex rotation method, are shown in Table 5.

Table 5. Singlet and triplet S-wave resonances in Ps^-. Positions are with respect to the ground state of the positronium. Units Are Ev.

n	Position	Width	Position	Width
	Singlet states		Triplet states	
2	4.7340	1.17(−3)	5.0742	1.36(−4)
	5.0709	2.74(−4)		
3	7.7646	2.04(−3)	6.0038	2.72(−4)
	5.9908	1.50(−3)		
4	6.2526	3.27(−3)	6.3383	2.72(−4)
	6.3267	4.08(−3)		
	6.3317	4.63(−3)		
5	6.4519	6.12(−3)		
	6.4723	1.91(−3)		

There are several doubly-excited or Feshbach-type triplet P even parity resonances below n = 2, 3, 4, 5, and 6 thresholds. These have been calculated by Ho and Bhatia [14]. They are given below each threshold in Table 6. The third resonance below n = 4 threshold is a shape resonance because it is above the n threshold.

Table 6. Doubly-excited $^3P^e$ resonances in Ps^-.

n	Position (Ry)	Width (Ry)
2	−0.12440 [a]	0.00054
3	−0.063261	3.58(−4)
	−0.0562095	5.78(−5)
4	−0.037789236	3.01(−5)
	−0.0033087	1.8(−6)
	0−0.030972 [a]	6.4(−5)
5	−0.02493166	5.09(−5)
	−0.0220972	5.24(−5)
	−0.021660	2.64(−5)
6	−0.017596	1.06(−4)
	−0.015894	1.6(−4)
	−0.015811	1.14(−4)
	−0.013761	4.0(−5)

[a] Shape resonance, above the threshold.

There are also odd parity triplet and singlet states which have been calculated by Bhatia and Ho [14–16]. Their results are shown in Tables 7 and 8. The lowest odd parity state in Ps^- has been observed by Michishio et al. [17] using laser beams of 2285 and 2297 angstroms and their results for the position and width agree with those given in Table 8.

Table 7. Odd parity triplet P-shape resonances of Ps^-.

n	State	Position (Ry)	Width (Ry)
3	^{3}P	−0.05450	9.20(−4)
5	^{3}P	−0.01971	6.60(−5)
7	^{3}P	−0.01008	4.00(−5)

Table 8. Odd parity singlet P-shape resonances of Ps^-.

n	Position (Ry)	Width (Ry)
2	−0.12434	9.00(−4)
4	−0.030975	6.00(−5)
6	−0.01375	5.20(−5)

Similar calculations have been carried out by Bhatia and Ho [18] for odd parity singlet and triplet D states of Ps^-. Their results are given in Table 9.

Table 9. Triplet and singlet D states of Ps^-. Positions are with respect to the ground state of Ps = −0.5 Ry.

N	E(Ry)	Width (Ry)	E(Ry)	Width (Ry)
	Triplet States		Singlet states	
3	−0.05589808	3.12(−6)	−0.060088253	1.43(−4)
4	−0.034501270	3.61(−4)	−0.032722544	2.60(−5)
	−0.03269579	3.10(−5)	−0.037098862	4.78(−5)
			−0.032722544	2.60(−4)
			−0.0302010 [a]	8.54(−4)
5	−0.023408248	1.75(−4)	−0.2467214	4.60(−5)
	−0.01977681 [a]	8.14(−5)	−0.2191458	4.47(−5)
			−0.02135816	2.30(−4)

[a] Shape resonance.

5. Resonances in PsH

As it is, it appears that the system having two neutral atoms, cannot have bound states. However, if it is viewed as a system consisting of a positron and H^-, then the positron is in the field of the Coulomb field of the hydrogen ion. There should be infinite number of bound states and resonances. Drachman and Houston [19] and Ho [20] have calculated the parameters of the S-wave resonance given below in Rydberg units:

$$E_R = -1.1726 \pm 0.0007 \; \Gamma = (4.6 \pm 1.1) \times 10^{-6} \tag{11}$$

by Drachman and Houston [19], and:

$$E_R = -1.205 \pm 0.001 \Gamma = (5.5 \pm 2.0) \times 10^{-3} \tag{12}$$

by Ho [20].

Similarly, there are P-wave and D-wave resonances.

A hybrid theory [21] has been developed, which considers the long-range and short-range correlations at the same time. The theory is variationally correct. In the scattering calculations, calculated phase shifts have lower bounds to the exact phase shifts. This method has been applied to calculate resonance parameters in He and Li^+. Phase shifts are calculated in the resonance region [22] and then fitted to the Breit–Wigner formula (cf. Equation (17) in [22]) to infer resonance parameters (cf. Equation (17) of [22]). The results obtained agree with those obtained using the Feshbach formalism [3]. The hybrid theory can also be applied to calculate resonance parameters in positron-target systems. Such calculations have not yet been carried out.

The books mentioned in references [1,23] have several chapters describing various methods employed to calculate resonance parameters. The author has two chapters in the book mentioned in ref. [1] (written with Aaron Temkin) on methods of calculating resonance parameters, also in [3].

6. Conclusions

Resonances in electron-target systems can be calculated easily. However, in the case of positron-target systems, calculations are not easy because of the positronium channel which is present below every threshold. As indicated above, correlations are very important. Therefore, wave functions having many terms are required. There are other systems with positrons where resonance parameters have been calculated. Here, we have discussed a few such systems and have described some of the methods of calculations. A theory called hybrid theory which includes the long-range and short-range correlations and is variationally correct has given resonance parameters for He and Li^+, which agree with those obtained in earlier calculations. This is achieved by calculating phase shifts in the resonance region and fitting them to the Breit–Wigner expression to infer resonance parameters. This approach could be applied to positron-target systems. There could be results in the future, using this approach.

Funding: This research received no external funding.

Conflicts of Interest: The author declares no conflict of interest.

References

1. Doschek, G.A. *Autoionization, Recent Developments and Applications*; Aaron, T., Ed.; Plenum Press: New York, NY, USA, 1985; p. 171.
2. Feshbach, H. Unified theory of nuclear reactions. *Ann. Phys. (N. Y.)* **1962**, *19*, 287. [CrossRef]
3. Bhatia, A.K.; Temkin, A. Calculation of autoionization of He and H^- using projection-operator formalism. *Phys. Rev. A* **1975**, *11*, 2018. [CrossRef]

4. Balslev, E.B.; Combes, J.W. Spectral properties of many-body Schrödinger Operators with Dialation-analytic interactions. Comm. *Math. Phys.* **1971**, *22*, 280. [CrossRef]
5. Mittleman, M.H. Resonances in Proton-Hydrogen and Positron-hydrogen. *Phys. Rev.* **1966**, *152*, 76. [CrossRef]
6. Doolan, G.D.; Nuttal, J.; Wherry, C.J. Evidence of a Resonance in a e^+-H S-wave scattering. *Phys. Rev. Lett.* **1978**, *40*, 313. [CrossRef]
7. Varga, K.; Mitroy, J.; Mezei, J.Z.; Kruppa, A.T. Description of positron-hydrogen resonances using the stockastic variational method. *Phys. Rev.* **2008**, *77*, 044502. [CrossRef]
8. Temkin, A.; Walker, J.F. Autoionization States of H^- below the n=2 Level of Hydrogen. *Phys. Rev.* **1965**, *140*, 1520. [CrossRef]
9. Bhatia, A.K.; Drachman, R.J. Search for resonances in positron-atom systems. *Phys. Rev.* **1990**, *42*, 5117. [CrossRef]
10. Ho, Y.K. Resonances in e^+-He^+ Scattering. *Phys. Rev. A* **1996**, *53*, 3165. [CrossRef]
11. Kar, S.; Ho, Y.K. S-wave resonances in e^+-He scattering below the Ps (n=2) excitation threshold. *J. Phys. B* **2004**, *37*, 3177. [CrossRef]
12. Ren, Z.; Han, H.; Shi, T. S-wave resonances in positron-helium scattering. *J. Phys. B* **2011**, *44*, 065204. [CrossRef]
13. Ho, Y.K. Doubly excited resonances of positronium negative ion. *Phys. Rev. A* **1984**, *102*, 348. [CrossRef]
14. Ho, Y.K.; Bhatia, A.K. Doubly excited $^3P^e$ resonant states in Ps^-. *Phys. Rev. A* **1992**, *45*, 6268. [CrossRef] [PubMed]
15. Bhatia, A.K.; Ho, Y.K. Complex-coordinate calculation of $^{1,3}P$ resonances in Ps^- using Hylleraas functions. *Phys. Rev. A* **1990**, *42*, 1119. [CrossRef] [PubMed]
16. Ho, Y.K.; Bhatia, A.K. P-wave shape resonances in positronium ions. *Phys. Rev. A* **1993**, *47*, 1497. [CrossRef] [PubMed]
17. Michishio, K.; Katani, T.; Kuma, S.; Azuma, T.; Wade, K.; Mochizuka, I.; Hydo, T.; Yogishita, A. Nagashima, Y. Observation of a shape resonance. *Nat. Artic.* **2016**, 11060. [CrossRef]
18. Ho, Y.K.; Bhatia, A.K. $^{1,3}D^o$ states in Ps^-. *Phys. Rev. A* **1994**, *50*, 2155. [CrossRef]
19. Drachman, R.J. Houston, Positronium-hydrogen elastic scattering. *Phys. Rev. A* **1975**, *12*, 885. [CrossRef]
20. Ho, Y.K. A resonant state and ground state of positronium hydride. *Phys. Rev. A* **1975**, *17*, 1675. [CrossRef]
21. Bhatia, A.K. Hybrid theory of electron-hydrogen scattering. *Phys. Rev. A* **2007**, *75*, 032713. [CrossRef]
22. Bhatia, A.K. Applications of the hybrid theory to the scattering of electrons from He^+ and Li^{2+} and resonances in these systems. *Phys. Rev. A* **2008**, *77*, 052707. [CrossRef]
23. Temkin, A. *Autoionization, Astrophysical, Theoretical, and Laboratory Experimental Aspects*; Aaron, T., Ed.; Mono Book Corp: Baltimore, MD, USA, 1966.

Review

Positron Processes in the Sun

Nat Gopalswamy

Solar Physics Laboratory, NASA Goddard Space Flight Center, Greenbelt, MD 20771, USA; nat.gopalswamy@nasa.gov

Received: 19 March 2020; Accepted: 16 April 2020; Published: 22 April 2020

Abstract: Positrons play a major role in the emission of solar gamma-rays at energies from a few hundred keV to >1 GeV. Although the processes leading to positron production in the solar atmosphere are well known, the origin of the underlying energetic particles that interact with the ambient particles is poorly understood. With the aim of understanding the full gamma-ray spectrum of the Sun, I review the key emission mechanisms that contribute to the observed gamma-ray spectrum, focusing on the ones involving positrons. In particular, I review the processes involved in the 0.511 MeV positron annihilation line and the positronium continuum emissions at low energies, and the pion continuum emission at high energies in solar eruptions. It is thought that particles accelerated at the flare reconnection and at the shock driven by coronal mass ejections are responsible for the observed gamma-ray features. Based on some recent developments I suggest that energetic particles from both mechanisms may contribute to the observed gamma-ray spectrum in the impulsive phase, while the shock mechanism is responsible for the extended phase.

Keywords: solar flares; coronal mass ejections; shocks; positrons; positronium; positron annihilation; pion decay

1. Introduction

Positrons, the antiparticle of electrons, were proposed theoretically by Dirac and were first detected by Anderson in 1933 [1]. Positrons are extensively used in the laboratory for a myriad of purposes (see review by Mills [2]). Astrophysical processes involving positrons have been found in the interstellar medium [3], and galactic bulge and disk [4]. Positrons are commonly found in the Sun [5]. The proton–proton chain, which accounts for most of the energy release inside the Sun involves the emission of a positron when two protons collide to form a deuterium nucleus. There are plenty of electrons present in the solar core, so the positrons are immediately annihilated with electrons and produce two gamma-ray photons. In one proton–proton chain, two positrons are emitted and hence contribute a total of four photons in addition to the two photons emitted during the formation of a helium-3 (^{3}He) nucleus from the fusion of deuterium nucleus with a proton.

High-energy particles exist in the solar atmosphere, energized during solar eruptions. Solar eruptions involve flares and coronal mass ejections (CMEs). A process known as magnetic reconnection taking place in the solar corona is thought to be the process which converts energy stored in the stressed solar magnetic fields into solar eruptions (see Figure 1 for a schematic of a typical eruption). One part of the released energy heats the plasma in the eruption region, while another goes to energize electrons and ions. Electromagnetic radiation from radio waves to gamma-rays produced by the energized electrons and protons by various processes is known as a flare. Acceleration in the reconnection region is referred to as flare acceleration or stochastic acceleration. CMEs also carry the released energy as the kinetic energy of the expelled magnetized plasma with a mass as high as 10^{16} g and speeds exceeding 3000 km/s. Such fast CMEs drive fast-mode magnetohydrodynamic shocks that can also energize ambient electrons and ions to very high energies. Diffusive shock acceleration and shock drift acceleration are the mechanisms by which particles are energized at the shock (see [7] for a review on acceleration mechanisms during solar eruptions). The flare and shock accelerations were referred to as first- and second-phase accelerations during an eruption by Wild et al. [8]. Accelerated particles have access to open and closed magnetic structures associated with the eruption resulting in a number of electromagnetic emissions via different emission mechanisms. Energetic particles escaping along interplanetary magnetic field lines are detected as solar energetic particle (SEP) events by particle detectors in space and on ground. These particles, originally known as solar cosmic rays, were first detected by Forbush [9] in the 1940s.

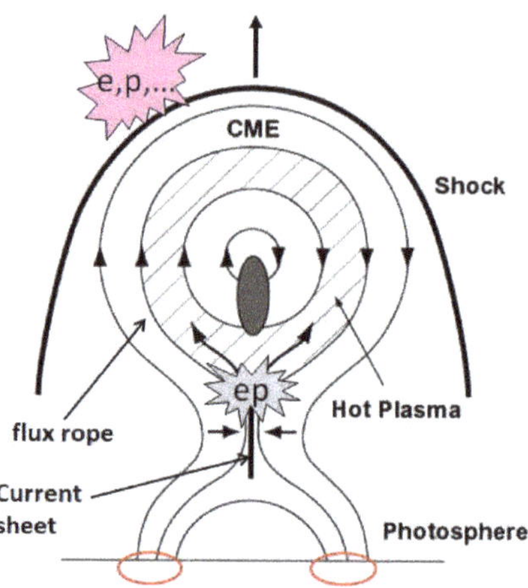

Figure 1. Schematic of a solar eruption and the sites of particle acceleration (e,p, . . .): one in the current sheet formed low in the corona and the other on the surface of the shock driven by the coronal mass ejection (CME) flux rope. The arrows toward the current sheet indicate the reconnection inflow while the ones diverging away indicate the outflow. The red ellipses in the photosphere represent the feet of flare loops where accelerated particles precipitate and produce flare radiation. Particles from the shock propagate away from the Sun and are detected as energetic particle events in space. The dark ellipse inside the flux rope represents a prominence that erupted along with the CME (adapted from Gopalswamy [6]).

Invoking the copious production of energetic particles from the Sun, Morrison [10] suggested that the active Sun must be a source of gamma-rays. He listed electron-positron annihilation as one of the processes expected to produce gamma-ray emission from cosmic sources [10]. Elliot [11] suggested that "positive electrons" from muon decay should lead to detectable 0.5 MeV gamma-ray line emission. Lingenfelter and Ramaty [12] performed detailed calculations of gamma-ray emission processes from the Sun. Chupp et al. [13] identified for the first time the positron annihilation radiation at 0.5 MeV along with other nuclear lines during the intense solar flares of 1972 August 4 and 7 using data from the Gamma-ray Monitor onboard NASA's Seventh Orbiting Solar Observatory (OSO-7) mission.

2. Mechanisms of Positron Production

Positrons are predominantly produced by three processes: (i) emission from radioactive nuclei, (ii) pair production by nuclear deexcitation, and (iii) decay of positively charged pions (π^+) that take place when ions accelerated in the corona interact with the ions in the photosphere/chromosphere. Kozlovsky et al. [14] list 156 positron-emitting radioactive nuclei resulting from the interaction of protons and α-particles (helium nuclei) with 12 different elements and their isotopes. The most important positron-emitting radioactive nuclei that result from the interaction of protons (p) and α-particles with carbon (^{12}C, ^{13}C), nitrogen (^{14}N, ^{15}N), and oxygen (^{16}O, ^{18}O) are listed in Table 1. In the interaction of p and α with ^{16}O, the oxygen nucleus, ends up in the excited state ($^{16}O^*$) of 6.052 MeV;

this nucleus deexcites by emitting an electron-positron pair with a lifetime of 0.096 ns. Another such excited nucleus is $^{40}Ca^*$ with a lifetime of 3.1 ns. Kozlovsky et al. [15] list another set of 23 positron emitters produced when accelerated ^3He interact with targets such as ^{12}C, ^{14}N, ^{16}O, ^{20}Ne, ^{24}Mg, ^{28}Si, and ^{56}Fe. The radioactive nuclei have a lifetime ranging from a tenth of a nanosecond to ~1 million years (see [14] for a list). There are 26 radioactive nuclei with a lifetime ≲10s.

Table 1. Important interactions and the resulting radioactive nuclei.

Interaction	Target Nuclei	Radioactive Nuclei
p - carbon	12C	11C, 12N, 10C, 13N
p - carbon	13C	13N
α - carbon	12C	11C, 15O
p - nitrogen	14N	11C, 13N, 14O
p - nitrogen	15N	15O
α - nitrogen	14N	17F, 13N, 11C
p - oxygen	16O	11C, 13N, 15O, 16O*
p - oxygen	18O	18F
α - oxygen	16O	19Ne, 18F, 15O, 13N, 11C, 16O*

Note: *Nucleus in excited state.

Positrons from radioactive nuclei have an energy of several hundred keV, while those from π^+ decay have much higher energy (up to hundreds of MeV). Almost all the positrons emitted by radioactive nuclei and a major fraction of those produced by π^+ decay (~80%) slow down to thermal levels (tens of eV) before directly annihilating or forming a positronium (Ps) atom by capturing an electron. The formation of Ps in this way is via radiative recombination. Ps can also be formed due to charge exchange with H and He atoms. Positronium is a hydrogen-like atom (with the proton replaced by a positron). There are two types of Ps, known as orthopositronium (O-Ps) and parapositromium (P-Ps), depending on how the spins of the positron and electron are oriented. In O-Ps, the electron and positron spins are in the same direction (triplet state); in P-Ps, the spins are oppositely directed (singlet state). O-Ps and P-Ps decay into 3 and 2 photons, respectively. O-Ps is formed preferentially: 75% of the time compared to 25% of the time for P-Ps [16]. Four key processes that determine the fate of the positrons produced in the solar atmosphere are discussed in [5]. These processes involve the interaction of positrons with the ambient hydrogen and helium in redistributing their energy, formation and quenching of Ps, and the ultimate production of gamma-rays by direct annihilation or via Ps. Direct annihilation of positrons can occur with free and bound (in H and He) electrons in the ambient medium. Positronium quenching occurs resulting in the emission of second-generation positrons when Ps collides with electrons, H, H^+, and He^+. Another quenching process is the conversion of O-Ps to P-Ps when O-Ps collides with electrons and H.

Pions (π^0 and $\pi^\pm$) are created when accelerated protons and α-particles from the corona collide with those in the chromosphere/photosphere. A detailed list of possible interactions (p-p and p-α) are listed in Murphy et al. [17]. High-energy positrons are primarily emitted from the decay of π^+ into positive muons (μ^+), which decay into positrons. In a similar reaction, negative pions (π^-) decay into negative muons (μ^-), which decay into electrons. π^0 decays into 2 gamma-rays most of the time (98.8%). In the remaining 1.2% of cases, π^0 decays into an electron positron pair and a gamma-ray. The rest energy of neutral pions is 135 MeV, while that of charged pions is 139.6 MeV. To produce these particles, the accelerated protons need to have high energies, exceeding ~300 MeV. The pions are very short-lived (π^0: 10^{-16} s; $\pi^\pm$: 2.6×10^{-8} s), while the muons live for a couple of microseconds (2.2×10^{-6} s). The >300 MeV protons needed for pion production seem to be accelerated both in the flare reconnection and CME-driven shocks (see Figure 1).

3. Gamma-Rays Due to Positrons

Nonthermal electromagnetic emission produced by charged particles is one of the key evidences for particle acceleration in the Sun (see, e.g., [18] for a review). Energetic electrons are readily inferred from nonthermal emission they produce at wavelengths ranging from millimeters to kilometers. Theories of radio emission have helped us understand the acceleration mechanism for the electrons and the radio emission mechanism. Energetic electrons also produce hard X-rays and gamma-rays. On the other hand, the electromagnetic indicators of energetic ions are limited to gamma-rays, ranging from a few hundred keV to >1 GeV. Positrons, which are produced by various processes noted in Section 2, contribute to the gamma-ray emission from the Sun at various energies via different processes.

Figure 2 shows the total gamma-ray spectrum from the Sun, exhibiting both line emission and continuum as predicted [10]. This spectrum is constructed from all possible processes that emit gamma-rays during solar eruptions [19]. At low energies (<10 MeV), there are many lines superposed on the electron bremsstrahlung continuum, the lowest being the 0.511 MeV line (positron annihilation, marked e^+). The next narrow line is the neutron capture line at 2.223 MeV. When energetic protons spallate ambient nuclei, neutrons are produced and emitted over a broad angular distribution; the downward neutrons slow down and are captured by a proton in the ambient medium forming a deuterium nucleus and releasing the binding energy as the 2.223 MeV line. The other lines are due to nuclear deexcitation of varying widths due to various combinations of incumbent and insurgent ions. The continuum emission has several components. Below the 0.511 MeV line, there is a weak Ps continuum that merges with the strong primary electron bremsstrahlung continuum. Primary electrons are those arriving from the acceleration site in the corona, as opposed to secondary electrons, which are produced in the chromosphere/photosphere due to the impact of accelerated ions from the corona (e.g., via π^- decay). The quasi-continuum between 0.7 MeV and 10 MeV (on which the discrete lines are superposed) is due to a multitude of broad nuclear lines caused by insurgent heavy ions interacting with ambient H and He nuclei. If the primary electron bremsstrahlung continuum is hard, it can be detected above background even at energies exceeding 10 MeV. When there is significant pion production both a broad line like feature centered near 70 MeV from neutral pions and a hard, secondary positron bremsstrahlung continuum can also be detected above 10 MeV (see below).

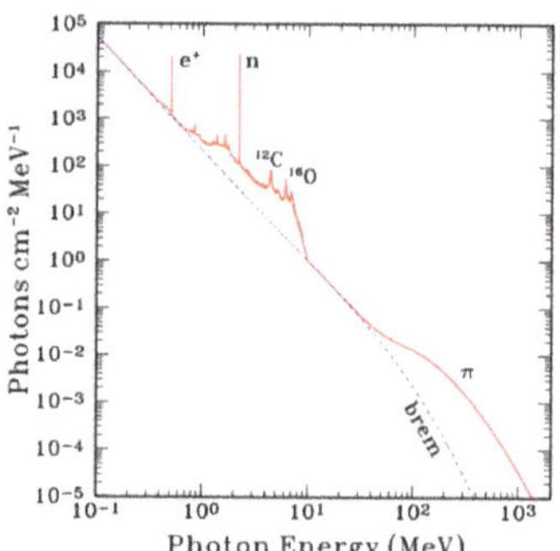

Figure 2. Overall theoretical spectrum of gamma-ray emission from the Sun from 0.1 MeV to 2 GeV. The blue dashed line labeled "brem" represents the contribution from the bremsstrahlung of energetic electrons accelerated during solar eruptions. The red line represents computed spectrum taking into account of all possible processes that contribute to gamma-ray emission. At energies below ~10 MeV there is a quasi-continuum with several lines superposed. e+ and n represent the 0.511 MeV positron annihilation line and the 2.223 MeV neutron capture line. ^{12}C and ^{16}O mark the next intense lines produced by nuclear deexcitation. The pion continuum at high energies is denoted by π, which involves contribution from both neutral and charged pions. The energy for these emission components is from energetic particles (electrons and ions) accelerated in the solar corona during solar eruptions (adapted from Ramaty and Mandzhavidze [19]).

3.1. The 0.511 MeV Gamma-Ray Line

The width of the solar annihilation line can range from ~1 keV to 10 keV full width at half maximum (FWHM) depending on the ambient conditions of the medium in which the annihilation takes place: e.g., temperature, density, and the ionization state [20]. Detailed calculations and comparison with observations have confirmed that the 0.511 MeV line is produced over a range of these parameters in the chromosphere/photosphere region [5]. One of the major contributors to the line width is the temperature in the region of annihilation because the ambient electrons have a distribution of speeds at a given temperature. The environmental conditions also determine the formation and destruction of Ps. Parapositronium annihilates emitting two 0.511 MeV photons (2γ decay), with lifetime ~0.125 ns. On the other hand, orthopositronium annihilates in three 341 keV photons ((3γ decay) with a lifetime ~142 ns [21]. Depending on the initial energy of the positron capturing an electron, the 3γ decay results in a gamma-ray continuum at energies below the annihilation line (see Figure 2). The flux ratio of the 3γ continuum (from O-Ps) to the 2γ line (from direct annihilation with free and bound electrons, and the decay of P-Ps) is an important parameter that can be used to infer the properties of the ambient medium (density, temperature, and ionization state). Murphy et al. [5] found that the flux ratio at a given temperature in an ionized medium remains constant up to an ambient hydrogen density of ~10^{13} cm^{-3} and then rolls over to values lower by 2–3 orders of magnitude at densities ~10^{17} cm^{-3}. The constant value depends on the ambient temperature starting from ~3 for 2000 K and decreasing to ~0.001 at 10 MK. For a neutral atmosphere, the temperature dependence is weak: the constant value of the flux ratio is ~5 at densities below ~10^{14} cm^{-3} and rolling over to ~0.008 at an ambient density of ~10^{17} cm^{-3}.

The 0.511 MeV emission can originate from different environments at different times during an event. Figure 3 shows the profile of the 0.511 MeV line during the 2003 October 28 eruption, considered to be an extreme event. In this event, the profile was quite wide during the first 2 min of the event compared to the last 16 min. Detailed calculations by Murphy et al. [5] revealed that the early part of the gamma-ray emission (broad line) might have occurred in an environment with a temperature in the range (3–4) × 10^5 K and densities $\gtrsim 10^{15}$ cm^{-3}. This implies that temperature in the chromosphere has transition-region values. Even though a 6000-K Vernazza et al. [22] model could marginally fit the observations, it was ruled out based on other considerations such as the low atmospheric density inferred (~2 × 10^{13} cm^{-3}). On the other hand, the narrow line late in the event is consistent with an environment in which the temperature is very low (~5000 K), the density is same as before (~10^{15} cm^{-3}), but the ionization fraction in the gas is ~20%. These results point to the inhomogeneous and dynamic nature of the chromosphere inferred from other considerations [23]. The derived conditions also depend on the atmospheric model, which itself has been revised [24].

3.2. Pion Continuum

The pion continuum described briefly above is shown in Figure 4 with different components: (i) the π^0 decay continuum, which has a characteristic peak around 68 MeV, (ii) the bremsstrahlung continuum due to positrons emitted by μ^+ resulting from π^+ decay (π^+ brm), (iii) the positron annihilation continuum due to in-flight annihilation of positrons from π^+ decay (π^+ ann), and (iv) the bremsstrahlung continuum due to electrons emitted by μ^- resulting from π^- decay (π^- brm). The sum of the four components (Total) represents the spectrum of gamma-rays resulting from pion decay assuming that the accelerated particle angular distribution is isotropic. The π^0 continuum dominates at energies >100 MeV and determines the spectrum at these energies but its contribution is very tiny at 10 MeV. For example, the π^- bremsstrahlung has four times larger contribution than from π^0 decay, while π^+ bremsstrahlung and annihilation contributions are larger by factors of 20 and 75, respectively. Below 10 MeV, the gamma-ray spectrum is mostly determined by π^+ bremsstrahlung. Thus, below ~30 MeV, the combined contribution from positrons dominate the spectrum. The π^0 continuum exceeds the π^+ bremsstrahlung around 30 MeV, producing the characteristic "giraffe" shoulder around this energy. Such a spectrum was first derived from the observations of pion continuum in the 1982 June 3 event by Forrest et al. [25], who identified a gamma-ray emission component that lasted for ~20 min beyond the impulsive phase of the flare. The spectrum in Figure 4 was calculated in great detail by Murphy et al. [17] to explain the 1982 June 3 event assuming that the primary protons have a shock spectrum [26]. They also performed a similar calculation assuming a stochastic particle spectrum thought to be produced in the reconnection site. While the early part of the 1982 June 3 gamma-rays can be explained by the steep stochastic spectrum, the part extending beyond the impulsive phase (late part) needs to be explained by the shock spectrum, which is much harder than the stochastic spectrum. Furthermore, these authors found that the shock spectrum is similar to the SEP spectrum observed in space.

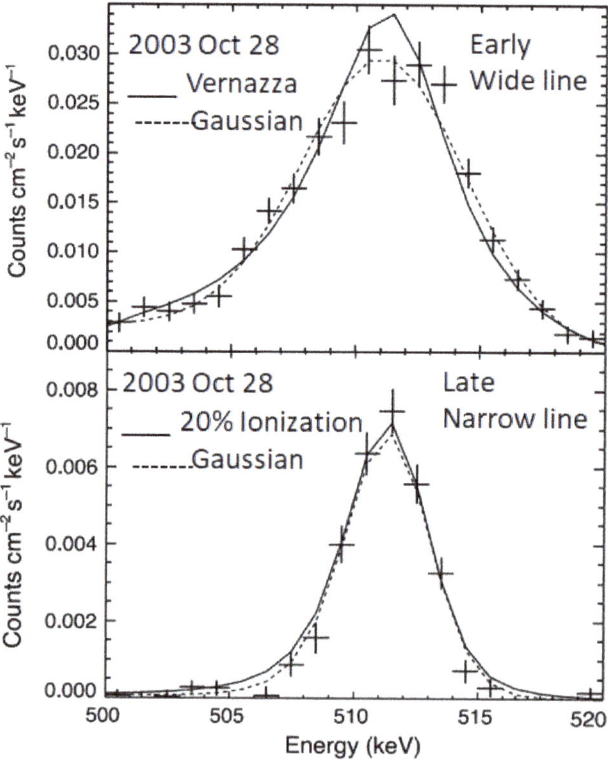

Figure 3. Profiles of the 0.511 MeV annihilation line during the 2003 October 28 event observed by the Reuven Ramaty High Energy Solar Spectroscopic Imager (RHESSI). (Top) During the early phase (first 2 min of observation) the data points are fitted with a Vernazza atmosphere [22] at a temperature of 6000 K and a Gaussian with a width of ~6.7 keV corresponding to a temperature of (3–4) × 10^5 K. (Bottom) During the late phase (last 16 min of observation), the data are fitted with a Gaussian (width ~1 keV) and a 5000-K atmosphere with 20% ionization (adapted from Murphy et al. [5]).

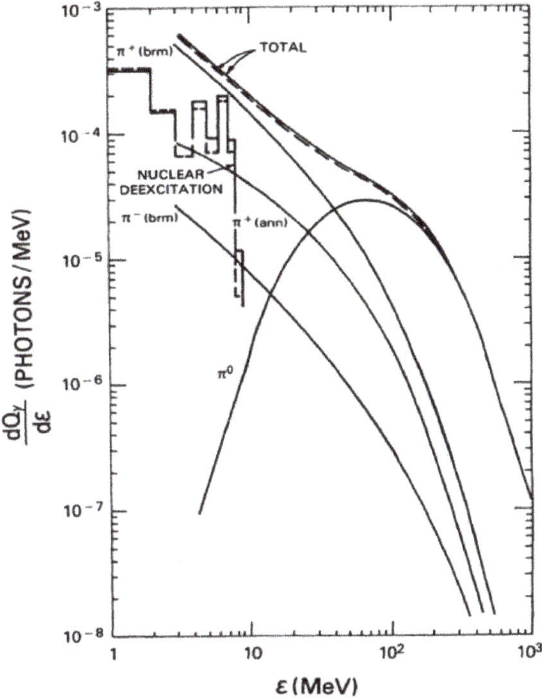

Figure 4. Contributions (in units of photons per unit energy interval) to the solar gamma-ray spectrum from π^0 decay, bremsstrahlung due to electrons from π^- decay (π^- brm), bremsstrahlung due to positrons from π^+ decay (π^+ brm), annihilation radiation of positrons from π^+ decay (π^+ ann). The energetic particles responsible for the production of pions were taken to be accelerated from a shock. The top two curves (solid and dashed) represent the total "giraffe" spectrum that combines these four components. The solid curve is for the ambient atmosphere while the dashed curve is for the abundance of the 1982 June 3 event derived from the associated solar energetic particle (SEP) event. The nuclear deexcitation line spectrum for the two abundances are also superposed. (Adapted from Murphy et al. [17]).

The time-extended gamma-ray emission first detected by Forrest et al. [25] using the Gamma-Ray Spectrometer on board the Solar Maximum Mission has been observed by many different missions, but such events were rare [27,28]. Two events had durations exceeding ~2 h [29,30]. The Large Area Telescope (LAT) on the Fermi satellite has detected dozens of such time-extended gamma-ray events from the Sun at energies >100 MeV, thanks to the detector's high sensitivity [31]. The average duration of these gamma-ray events is about 9.7 h and with six events lasting for more than 12 h [32,33], including an event that lasted for almost a day. The time-extended events are known by different names "long duration gamma-ray flare (LDGRF)" [34,35], "sustained gamma-ray emission (SGRE)" [33,36], and "late-phase gamma-ray emission (LPGRE)" [32]. The Fermi/LAT observations have revived the interest in the origin of the high-energy particles in these events because the accelerator needs to inject >300 MeV ions toward the chromosphere/photosphere to produce the pions required for the gamma-ray events.

Works focusing on the time-extended nature of these gamma-ray events explore ways to extend the life of the >300 MeV protons from stochastic (impulsive-phase) acceleration, e.g., by particle trapping in flare loops (e.g., [37]). In this scenario, the largest spatial extent of the gamma-ray source is the size of the post-eruption arcade (or flare area) discerned from coronal images taken in extreme ultraviolet wavelengths. In the shock scenario, the gamma-ray source is spatially extended because the angular extent of the shock is much larger than that of the flare structure [38,39]; shock acceleration is naturally time-extended evidenced by type II radio bursts and SEP events [33]

Gopalswamy et al. [40] demonstrated the spatially extended nature of the gamma-ray source during the 2014 September 1 event (see also [36,39,41]). They used multiview data from the Solar and Heliospheric Observatory (SOHO) and the Solar Terrestrial Relations Observatory (STEREO) missions to obtain a detailed picture of the eruption, including magnetic structures that extend beyond the flare structure (post-eruption arcade, PEA) as described in Figure 5. It must be noted that both the flux rope and PEA are products of the eruption process (magnetic reconnection). The PEA remains anchored to the solar surface, while the flux rope is ejected into the heliosphere with speeds exceeding 2000 km/s. The flux rope is a large structure rapidly expanding into the heliosphere compared to the compact flare structure. The flux rope drives a shock because of its high speed and the shock accelerates the required >300 MeV protons. The protons travel down to the chromosphere/photosphere along the field lines located between the flux rope and shock and produce the gamma-rays. Particles traveling away from the shock into the heliosphere are detected as SEP events. Shocks are known to accelerate particles as they propagate into the heliosphere beyond Earth's orbit, but the high-energy particles required for pion production may be accelerated only to certain distance from the Sun; this distance determines the duration of an SGRE event. In the case of the 2014 September 1 event, the SGRE lasted for about 4 h. With a shock speed >2300 km/s obtained from coronagraph observation, one can infer that the shock stopped accelerating >300 MeV protons to sufficient numbers by the time it reached a distance of about 50 solar radii from the Sun. Evidence for the shock shown in Figure 5 is the interplanetary type II radio burst that lasted until the end of the SGRE event and a bit beyond. The estimated duration of the type II burst (about 7.5 h) is in agreement with the linear relation found between the two durations [33].

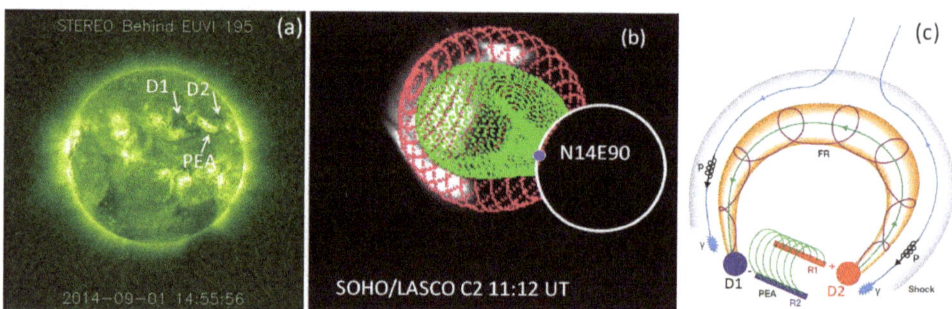

Figure 5. (**a**) An extreme ultraviolet image obtained by the STEREO mission showing the spatial structure of the eruption region consisting of dimming regions (D1, D2) and the post-eruption arcade (PEA). (**b**) A flux rope (green) and shock (red) structures superposed on a SOHO white-light image showing the CME. The blue dot (at heliographic coordinates N14E90) represents the centroid of the Fermi/LAT gamma-ray source located between the flux rope (FR) leg and the shock front. The flux rope legs are rooted in the dimming regions D1 and D2. (**c**) A schematic showing the FR and the surrounding shock. Particles accelerated near the shock nose travel along magnetic field lines in the space between the FR and shock, precipitate in the chromosphere/photosphere to produce gamma-rays via the pion-decay mechanisms discussed in the text (adapted from Gopalswamy et al. [40]).

In a given eruption, both flare and shock populations are expected to be present, the flare particles being the earliest. Murphy et al. [17] concluded that the nuclear deexcitation line flux is primarily due to the flare population, while the 0.511-line flux has roughly equal contributions from flare and shock populations. On the other hand, the extended phase emission is solely due to the shock population. The conclusion on the extended phase emission initially derived from studying the 1982 June 3 event, seems to be applicable to all events with time-extended emission [33]. Recently, Minasyants et al. [42]

found that in certain gamma-ray events with high fluxes of >100 MeV photons, the development of the flare and CME are simultaneous. The CME started during the impulsive phase of the flare. They found that the >100 MeV flux is highly correlated with the CME speed, although the sample is small. Figure 6 shows the relation, replotted with a second-order polynomial fit instead of their linear fit. Normally, one would have thought the impulsive phase gamma-ray flux should be related to the impulsive-phase proton population, and not to the shock population. On further examination, it is found that a type II radio burst started in the impulsive phase of the events, indicating early shock formation. This is typical of eruptions that produce ground level enhancement (GLE) in SEP events [43–45], implying particle acceleration by shocks to GeV energies within the impulsive phase. This result suggests that the shock population may also have contribution to >100 MeV photons in the impulsive phase.

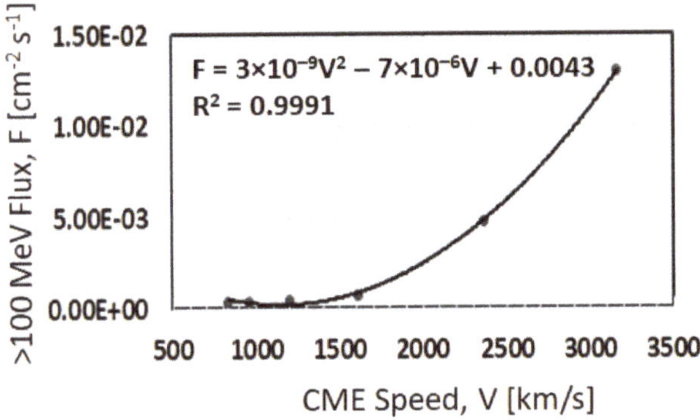

Figure 6. Scatter plot between >100 MeV gamma-ray flux (F) from Fermi/LAT against the CME speed (V) for events in which the CME onset was during the impulsive phase. The speeds of three CMEs are different from those in Minasyants et al. [42]. The polynomial fit to the data points and the square of the correlation coefficient (R) are shown on the plot.

STEREO observations have revealed that shocks can form very close to the Sun, as close as 1.2 solar radii [46]. Shocks typically take several minutes to accelerate particles to GeV energies after their formation. This means, shocks are present within the closed field regions of the corona early on, sending particles toward the Sun and augmenting the impulsive phase particles. The precipitation sites are expected to be different from the PEA as discussed in Figure 5. Once the shock propagates beyond ~2.5 solar radii, accelerated particles can move both ways, away and toward the Sun.

4. Conclusions

Positrons are important particles both in the laboratory and in astrophysics. They are extremely useful in understanding high-energy phenomena on the Sun. They provide information on various processes starting from particle acceleration, transport, and interaction with the dense part of the solar atmosphere. Positrons provide information on the physical conditions in the chromosphere/photosphere where they are produced and destroyed by different processes, leaving tell-tale signatures in the gamma-ray spectrum. In addition, gamma-rays and positrons provide information on the magnetic structures involved in solar eruptions that support the acceleration and transport of the highest-energy particles in the inner heliosphere. Spatially resolved gamma-ray observations beyond what is currently available (e.g., [47]) are needed to resolve the issue of the relative contributions from stochastic and shock accelerations in solar eruptive events.

Funding: This research was funded by NASA's Living with a Star Program.

Acknowledgments: I thank Gerald H. Share and Pertti A. Mäkelä for reading versions of this manuscript and providing useful comments.

Conflicts of Interest: The author declares no conflict of interest.

References

1. Anderson, C.D. The Positive Electron. *Phys. Rev.* **1933**, *43*, 491–494. [CrossRef]
2. Mills, A.P. 2-Positron and Positronium Sources. In *Methods in Experimental Physics*; Dunning, F.B., Hulet, R.G., Eds.; Atomic, Molecular, and Optical Physics: Charged Particles; Academic Press: Cambridge, MA, USA, 1995; Volume 29, pp. 39–68. [CrossRef]
3. Guessoum, N.; Jean, P.; Gillard, W. The Lives and Deaths of Positrons in the Interstellar Medium. *Astron. Astrophys.* **2005**, *436*, 171–185. [CrossRef]
4. Knödlseder, J.; Jean, P.; Lonjou, V.; Weidenspointner, G.; Guessoum, N.; Gillard, W.; Skinner, G.; von Ballmoos, P.; Vedrenne, G.; Roques, J.-P.; et al. The All-Sky Distribution of 511 KeV Electron-Positron Annihilation Emission. *Astron. Astrophys.* **2005**, *441*, 513–532. [CrossRef]
5. Murphy, R.J.; Share, G.H.; Skibo, J.G.; Kozlovsky, B. The Physics of Positron Annihilation in the Solar Atmosphere. *Astrophys. J. Suppl. Ser.* **2005**, *161*, 495–519. [CrossRef]
6. Gopalswamy, N. Properties of Interplanetary Coronal Mass Ejections. *Space Sci. Rev.* **2006**, *124*, 145–168. [CrossRef]
7. Aschwanden, M.J. Particle Acceleration and Kinematics in Solar Flares—A Synthesis of Recent Observations and Theoretical Concepts (Invited Review). *Space Sci. Rev.* **2002**, *101*, 1–227. [CrossRef]
8. Wild, J.P.; Smerd, S.F.; Weiss, A.A. Solar Bursts. *Annu. Rev. Astron. Astrophys.* **1963**, *1*, 291. [CrossRef]
9. Forbush, S.E. Three Unusual Cosmic-Ray Increases Possibly Due to Charged Particles from the Sun. *Phys. Rev.* **1946**, *70*, 771–772. [CrossRef]
10. Morrison, P. Solar Origin of Cosmic Ray Time Variations. *Il Nuovo Cimento* **1958**, *7*, 858–865. [CrossRef]
11. Elliot, H. The Nature of Solar Flares. *Planet. Space Sci.* **1964**, *12*, 657–660. [CrossRef]
12. Lingenfelter, R.E.; Ramaty, R. *High Energy Nuclear Reactions in Solar Flares*; Univrsity of California, Los Angeles: Los Angeles, CA, USA, 1967; pp. 99–124.
13. Chupp, E.L.; Forrest, D.J.; Higbie, P.R.; Suri, A.N.; Tsai, C.; Dunphy, P.P. Solar Gamma Ray Lines Observed during the Solar Activity of 2 August to 11 August 1972. *Nature* **1973**, *241*, 333–335. [CrossRef]
14. Kozlovsky, B.; Lingenfelter, R.E.; Ramaty, R. Positrons from Accelerated Particle Interactions. *Astrophys. J.* **1987**, *316*, 801–818. [CrossRef]
15. Kozlovsky, B.; Murphy, R.J.; Share, G.H. Positron-Emitter Production in Solar Flares from 3He Reactions. *Astrophys. J.* **2004**, *604*, 892–899. [CrossRef]
16. Ramaty, R.; Murphy, R.J. Nuclear Processes and Accelerated Particles in Solar Flares. *Space Sci. Rev.* **1987**, *45*, 213–268. [CrossRef]
17. Murphy, R.J.; Dermer, C.D.; Ramaty, R. High-Energy Processes in Solar Flares. *Astrophys. J. Suppl. Ser.* **1987**, *63*, 28. [CrossRef]
18. Benz, A.O. Flare Observations. *Living Rev. Sol. Phys.* **2016**, *14*, 2. [CrossRef]
19. Ramaty, R.; Mandzhavidze, N. Solar Flares: Gamma Rays. In *Encyclopedia of Astronomy and Astrophysics*; Murdin, P., Ed.; Article id. 2292; Institute of Physics Publishing: Bristol, UK, 2001; pp. 1–4. Available online: http://eaa.crcpress.com/default.asp (accessed on 22 April 2020).
20. Share, G.H.; Murphy, R.J.; Smith, D.M.; Schwartz, R.A.; Lin, R.P. RHESSI e+-e- Annihilation Radiation Observations: Implications for Conditions in the Flaring Solar Chromosphere. *Astrophys. J. Lett.* **2004**, *615*, L169–L172. [CrossRef]
21. Ley, R. Atomic Physics of Positronium with Intense Slow Positron Beams. *Appl. Surf. Sci.* **2002**, *194*, 301–306. [CrossRef]
22. Vernazza, J.E.; Avrett, E.H.; Loeser, R. Structure of the Solar Chromosphere. III - Models of the EUV Brightness Components of the Quiet-Sun. *Astrophys. J. Suppl. Ser.* **1981**, *45*, 635–725. [CrossRef]
23. Song, P. A Model of the Solar Chromosphere: Structure and Internal Circulation. *Astrophys. J.* **2017**, *846*, 92. [CrossRef]

24. Avrett, E.H.; Loeser, R. Models of the Solar Chromosphere and Transition Region from SUMER and HRTS Observations: Formation of the Extreme-Ultraviolet Spectrum of Hydrogen, Carbon, and Oxygen. *Astrophys. J. Suppl. Ser.* **2008**, *175*, 229–276. [CrossRef]
25. Forrest, D.J.; Vestrand, W.T.; Chupp, E.L.; Rieger, E.; Cooper, J.F.; Share, G.H. Neutral Pion Production in Solar Flares. *ICRC* **1985**, *4*, 146–149.
26. Ellison, D.C.; Ramaty, R. Shock Acceleration of Electrons and Ions in Solar Flares. *Astrophys. J.* **1985**, *298*, 400–408. [CrossRef]
27. Chupp, E.L.; Ryan, J.M. High Energy Neutron and Pion-Decay Gamma-Ray Emissions from Solar Flares. *Res. Astron. Astrophys.* **2009**, *9*, 11–40. [CrossRef]
28. Kuznetsov, S.N.; Kurt, V.G.; Yushkov, B.Y.; Myagkova, I.N.; Galkin, V.I.; Kudela, K. Protons Acceleration in Solar Flares: The Results of the Analysis of Gamma-Emission and Neutrons Recorded by the SONG Instrument Onboard the CORONAS-F Satellite. *Astrophys. Space Sci. Lib.* **2014**, *400*, 301. [CrossRef]
29. Akimov, V.V.; Afanassyey, V.G.; Belaousov, A.S.; Blokhintsev, I.D.; Kalinkin, L.F.; Leikov, N.G.; Nesterov, V.E.; Volsenskaya, V.A.; Galper, A.M.; Chesnokov, V.J.; et al. Observation of High Energy Gamma-Rays from the Sun with the GAMMA-1 Telescope (E > 30 MeV). *ICRC* **1991**, *3*, 73.
30. Kanbach, G.; Bertsch, D.L.; Fichtel, C.E.; Hartman, R.C.; Hunter, S.D.; Kniffen, D.A.; Kwok, P.W.; Lin, Y.C.; Mattox, J.R.; Mayer-Hasselwander, H.A. Detection of a Long-Duration Solar Gamma-Ray Flare on 11 June 1991 with EGRET on COMPTON-GRO. *Astron. Astrophys. Suppl. Ser.* **1993**, *97*, 349–353.
31. Atwood, W.B.; Abdo, A.A.; Ackermann, M.; Althouse, W.; Anderson, B.; Axelsson, M.; Baldini, L.; Ballet, J.; Band, D.L.; Barbiellini, G.; et al. The Large Area Telescope on the Fermi Gamma-Ray Space Telescope Mission. *Astrophys. J.* **2009**, *697*, 1071–1102. [CrossRef]
32. Share, G.H.; Murphy, R.J.; White, S.M.; Tolbert, A.K.; Dennis, B.R.; Schwartz, R.A.; Smart, D.F.; Shea, M.A. Characteristics of Late-Phase $\greater$100 MeV Gamma-Ray Emission in Solar Eruptive Events. *Astrophys. J.* **2018**, *869*, 182. [CrossRef]
33. Gopalswamy, N.; Mäkelä, P.; Yashiro, S.; Lara, A.; Akiyama, S.; Xie, H. On the Shock Source of Sustained Gamma-Ray Emission from the Sun. *J. Phys. Conf. Ser.* **2019**, *1332*, 012004. [CrossRef]
34. Ryan, J.M. Long-Duration Solar Gamma-Ray Flares. *Space Sci. Rev.* **2000**, *93*, 581–610. [CrossRef]
35. De Nolfo, G.A.; Bruno, A.; Ryan, J.M.; Dalla, S.; Giacalone, J.; Richardson, I.G.; Christian, E.R.; Stochaj, S.J.; Bazilevskaya, G.A.; Boezio, M.; et al. Comparing Long-Duration Gamma-Ray Flares and High-Energy Solar Energetic Particles. *Astrophys. J.* **2019**, *879*, 90. [CrossRef]
36. Plotnikov, I.; Rouillard, A.P.; Share, G.H. The Magnetic Connectivity of Coronal Shocks from Behind-the-Limb Flares to the Visible Solar Surface during γ-Ray Events. *Astron. Astrophys.* **2017**, *608*, A43. [CrossRef]
37. Ryan, J.M.; Lee, M.A. On the Transport and Acceleration of Solar Flare Particles in a Coronal Loop. *Astrophys. J.* **1991**, *368*, 316. [CrossRef]
38. Cliver, E.W.; Kahler, S.W.; Vestrand, W.T. On the Origin of Gamma-Ray Emission from the Behind-the-Limb Flare on 29 September 1989. *ICRC* **1993**, *3*, 91.
39. Pesce-Rollins, M.; Omodei, N.; Petrosian, V.; Liu, W.; Rubio da Costa, F.; Allafort, A.; Fermi-LAT Collaboration. Fermi Large Area Telescope Observations of High-Energy Gamma-Ray Emission from behind-the-Limb Solar Flares. *ICRC* **2015**, *34*, 128.
40. Gopalswamy, N.; Mäkelä, P.; Yashiro, S.; Akiyama, S.; Xie, H.; Thakur, N. Source of Energetic Protons in the 2014 September 1 Sustained Gamma-Ray Emission Event. *Sol. Phys.* **2020**, *295*, 18. [CrossRef]
41. Jin, M.; Petrosian, V.; Liu, W.; Nitta, N.V.; Omodei, N.; Rubio da Costa, F.; Effenberger, F.; Li, G.; Pesce-Rollins, M.; Allafort, A.; et al. Probing the Puzzle of Behind-the-Limb γ-Ray Flares: Data-Driven Simulations of Magnetic Connectivity and CME-Driven Shock Evolution. *Astrophys. J.* **2018**, *867*, 122. [CrossRef]
42. Minasyants, G.; Minasyants, T.; Tomozov, V. Features of Development of Sustained Fluxes of High-Energy Gamma-Ray Emission at Different Stages of Solar Flares. *Sol.-Terr. Phys.* **2019**, *5*, 10–17. [CrossRef]
43. Reames, D.V. Solar Energetic-Particle Release Times in Historic Ground-Level Events. *Astrophys. J.* **2009**, *706*, 844–850. [CrossRef]
44. Gopalswamy, N.; Xie, H.; Yashiro, S.; Akiyama, S.; Mäkelä, P.; Usoskin, I.G. Properties of Ground Level Enhancement Events and the Associated Solar Eruptions during Solar Cycle 23. *Space Sci. Rev.* **2012**, *171*, 23–60. [CrossRef]

45. Gopalswamy, N.; Xie, H.; Akiyama, S.; Yashiro, S.; Usoskin, I.G.; Davila, J.M. The First Ground Level Enhancement Event of Solar Cycle 24: Direct Observation of Shock Formation and Particle Release Heights. *Astrophys. J. Lett.* **2013**, *765*, L30. [CrossRef]
46. Gopalswamy, N.; Xie, H.; Mäkelä, P.; Yashiro, S.; Akiyama, S.; Uddin, W.; Srivastava, A.K.; Joshi, N.C.; Chandra, R.; Manoharan, P.K.; et al. Height of Shock Formation in the Solar Corona Inferred from Observations of Type II Radio Bursts and Coronal Mass Ejections. *Adv. Space Res.* **2013**, *51*, 1981–1989. [CrossRef]
47. Omodei, N.; Pesce-Rollins, M.; Longo, F.; Allafort, A.; Krucker, S. Fermi-LAT Observations of the 2017 September 10 Solar Flare. *Astrophys. J. Lett.* **2018**, *865*, L7. [CrossRef]

Article

Analytical Results for the Three-Body Radiative Attachment Rate Coefficient, with Application to the Positive Antihydrogen Ion $\overline{\text{H}}^+$

Jack C. Straton

Department of Physics, Portland State University, Portland, OR 97207-0751, USA; straton@pdx.edu

Received: 16 March 2020; Accepted: 10 April 2020; Published: 20 April 2020

Abstract: To overcome the numerical difficulties inherent in the Maxwell–Boltzmann integral of the velocity-weighted cross section that gives the radiative attachment rate coefficient α_{RA} for producing the negative hydrogen ion H^- or its antimatter equivalent, the positive antihydrogen ion $\overline{\text{H}}^+$, we found the analytic form for this integral. This procedure is useful for temperatures below 700 K, the region for which the production of $\overline{\text{H}}^+$ has potential use as an intermediate stage in the cooling of antihydrogen to ultra-cold (sub-mK) temperatures for spectroscopic studies and probing the gravitational interaction of the anti-atom. Our results, utilizing a 50-term explicitly correlated exponential wave function, confirm our prior numerical results.

Keywords: antihydrogen; radiative attachment; photodetachment; antihydrogen ion; analytical; hydrogen ion

---

1. Introduction

The Antiproton Decelerator (AD) facility at CERN [1] has provided the foundation for a variety of experiments (e.g., [2–4]) over more than a decade. Small numbers of these anti-atoms are trapped by the ALPHA and ATRAP collaborations using specialized magnetic minimum neutral atom traps [5–7], with confinement times of many minutes being routine at ALPHA [8]. They have done spectroscopic [9] measurements for $\overline{\text{H}}$ in their quest to investigate possible violations of CPT symmetry, experimental limits on its charge [10], and preliminary limits on the gravitational interaction of the anti-atom [11].

Building on the latter idea, the GBAR collaboration [12–14] means to measure the gravitational attraction of matter versus antimatter using neutral $\overline{\text{H}}$ atoms, but cooling them sufficiently is difficult because of their neutrality. They intend to form the antihydrogen ion $\overline{\text{H}}^+$ as an intermediate step because its net charge would allow for sympathetic cooling with a mixture of positively charged ions of ordinary matter such as Be^+, and, after they are cooled, the extra positron would be stripped off prior to studies of the gravitational interaction of the anti-atom [12–14]. The authors of [15,16] calculated the cross section and rate coefficient for the radiative attachment of a second positron to create the $\overline{\text{H}}^+$ ion,

$$\overline{\text{H}}\,(1s) + e^+ \to \overline{\text{H}}^+ \left(1s^2\,{}^1S^e\right) + \hbar\omega \quad . \tag{1}$$

We first [15] used the effective range wave function of Ohmura and Ohmura [17] and then [16] a fully two-positron 200-term wave function [18] composed of explicitly correlated exponentials of the kind introduced by Thakkar and Smith [19]. These extend to temperatures lower than Bhatia's [20] results.

Calculating the radiative attachment rate coefficient α_{RA} for producing the negative hydrogen ion H^- or its antimatter equivalent, the positive antihydrogen ion $\overline{\text{H}}^+$, requires the evaluation of a Maxwell–Boltzmann integral of the velocity-weighted cross section whose integrand is akin to a slightly-rounded Heaviside step function that is difficult to handle numerically, particularly for temperatures below 1000 K. Evaluating this integral analytically would, then, be ideal, perhaps using the analytical results for the underlying six-dimensional photoionization integral for the cross section itself given in Keating's master's thesis [21]. However, integrating squares of sums of the large variety of terms in that final cross section is a daunting task.

This variety of terms arises in Keating's work as the Lth derivatives of the Laplace transform of the spherical Bessel function $j_1(kr)$,

$$\mathcal{L}_L(p;k) = \begin{cases} \frac{1}{k^2}\left[k - p\,\tan^{-1}\left(\frac{k}{p}\right)\right] & L=0 \\ \frac{1}{k^2}\left[-\frac{kp}{p^2+k^2} + \tan^{-1}\left(\frac{k}{p}\right)\right] & L=1 \\ \frac{i(L-2)!}{2k^2}\left[\frac{p+ikL}{(p+ik)^L} - \frac{p-ikL}{(p-ik)^L}\right] = & \\ \dfrac{-(L-2)!}{k\left(k^2+p^2\right)^L}\sum_{j=0}^{L}{}^{(2)}\binom{L}{j}\dfrac{j(L+1)}{j+1}(-1)^{j/2}k^j p^{L-j} & L=2,3,4,\ldots \end{cases} \tag{2}$$

where the "(2)" on the summation sign in the last line indicates steps of two, $p = \alpha+\gamma$ or $\alpha+\beta$ of Equation (8), and k is the wave number. One might wonder, then, if one could back up to the final radial integral of the cross section that has a consistent analytic form $r^{3/2+h}e^{-\sigma r}j_1(kr)$ for the direct and cross terms. It indeed turns out to be possible to perform the Maxwell–Boltzmann integral of products of that analytic form first, and then integrate over each of the radial integrals in the product $r^{3/2+h}e^{-\sigma r}j_1(kr)\ R^{3/2+j}e^{-\tau r}j_1(kR)$, and finally sum over all such product terms and the terms of the explicitly correlated exponential wave function.

We give a synopsis of how one finds the radiative attachment cross section, explicate the integrals one needs to calculate, and show how a fully analytical rate coefficient may be found. Tests of the new form confirm the numerical integrals of prior work.

2. The Radiative Attachment cross Section

Since this approach relies on finding the radiative attachment cross section from the photodetachment cross section via the principle of detailed balance, we give a short history. Photodetachment from the hydrogen ion H^-, for instance, is known to be responsible for the opacity of the sun [22,23], garnering much attention in the 1940s–1980s [24–42] and more recently [43–51]. Ward, McDowell, and Humberston [52] described the parallel Ps^- case as calculating an allowed dipole transition to the continuum of the two-electron (or two-positron) Hamiltonian

$$\hbar\omega + \mathrm{H}^-\left(1s^2\,{}^1S_0^e\right) \rightarrow \mathrm{H}^-\left(1s\ \mathrm{kp}\,{}^1P_0^o\right). \tag{3}$$

Following Ghoshal and and Ho [50], we put the (length gauge[1]) cross section for photodetachment (or photoionization), σ_{PI}, laid out by Chandrasekhar [24,25] in atomic units, and obtain

$$\begin{aligned} \sigma_{PI} &= \frac{2p\omega\alpha a_0^2}{3}\left|\left\langle\psi_f\left|\hat{k}\cdot(\mathbf{r}_1+\mathbf{r}_2)\right|\psi_i\right\rangle\right|^2 \\ &= 6.81156\times10^{-20}\ \mathrm{cm}^2 k\left(k^2+2I\right)\left|\left\langle\psi_f\left|z_1+z_2\right|\psi_i\right\rangle\right|^2, \end{aligned} \tag{4}$$

where α is the fine structure constant, a_0 is the Bohr radius, and the magnitude of the momentum of the detached electron or positron $p = \hbar k$ may be related to the energy ω (and the photon wavelength λ) and the H^- electron or positron affinity, I, by

1 Since the cross section differences between velocity and length gauge formulations (due to the approximate nature of the two-positron wave functions used) are small, we will present only length gauge results in this work.

$$2\omega = 2\,(2\pi\nu) = 2\frac{hc}{\lambda} = 2\left(\frac{\hbar^2 k^2}{2} + I\right) \equiv \left(p^2 + \gamma_0^2\right). \tag{5}$$

In the matrix element for photodetachment

$$\mu_{PI} = \int \psi_f^*(z_1 + z_2)\psi_i d\tau \tag{6}$$

in Equation (4), ψ_f is a continuum state wave function for the outgoing positron represented by (the dipole term of) a plane wave multiplied by a hydrogen ground state wave function,

$$\psi_f = \frac{1}{\sqrt{2\pi}}(e^{ikz_1 - r_2} + e^{ikz_2 - r_1}) \tag{7}$$

Accurate two-positron initial-state wave functions may take the form

$$\psi_H(r_1, r_2, r_{12}) = \frac{1}{\sqrt{2}}\left(1 - \hat{P}_{12}\right) e^{-\alpha r_1 - \beta r_2 - \gamma r_{12}} \sum_{l,m,n} c_{lmn} s^l t^{2m} u^n, \tag{8}$$

where $\hat{P}_{12}$ is the permutation operator for the two identical positrons $\alpha \leftrightarrow \beta$, with Hylleraas coordinates [53] given by $s = r_1 + r_2$, $t = r_1 - r_2$, and $u = r_{12} \equiv |\mathbf{r}_1 - \mathbf{r}_2|$. One may also express this as sums of powers of r_1 and r_2 instead of powers of s and t, via the binomial theorem. Often, the difficulty of finding the nonlinear parameters in the exponential is reduced by setting $\beta = \alpha$ and $\gamma = 0$.

Alternatively, Thakkar and Smith [19] introduced a set of wave functions involving solely exponentials, with the nonlinear inter-electron (inter-positron) correlation parameter γ retained,

$$\psi_{TS}(r_1, r_2, r_{12}) = \frac{1}{\sqrt{2}}\left(1 - \hat{P}_{12}\right) \sum_k c_k e^{-\alpha_k r_1 - \beta_k r_2 - \gamma_k r_{12}}, \tag{9}$$

where the parameters in the exponentials are generated in a quasi-random fashion,

$$\begin{aligned} \alpha_k &= \eta\left((A_2 - A_1)\tfrac{1}{2}\langle k(k+1)\rangle\,\sqrt{2} + A_1\right) \\ \beta_k &= \eta\left((B_2 - B_1)\tfrac{1}{2}\langle k(k+1)\rangle\,\sqrt{3} + B_1\right), \\ \gamma_k &= \eta\left((G_2 - G_1)\tfrac{1}{2}\langle k(k+1)\rangle\,\sqrt{5} + G_1\right) \end{aligned} \tag{10}$$

where $\langle x\rangle$ denotes the fractional part of x. The downside of having to find six nonlinear parameters that minimize the energy, rather than the single nonlinear parameter one varies in a many Hylleraas expansions, is sufficiently compensated for in that the wave function has a consistent form and is generally easier for evaluating integrals. For the fifty-term wave function we use, these parameters are [54]: $A_1 = 0.2380$, $A_2 = 1.3240$, $B_1 = 0.9800$, $B_2 = 1.3290$, $G_1 = -0.0720$, $G_2 = 0.288$, and $\eta = 1 - 2.458 \times 10^{-7}$. The quasi-random assignment of the 50 values for each of α_k, β_k, and γ_k in Equation (10) means that we do not have to vary these 150 parameters directly. Because the optimization algorithm is not perfect, one must scale the wave function with η very slightly different from one so that it satisfies the virial theorem. The coefficients c_k are found by diagonalizing the Hamiltonian matrix in order to minimize the ground state energy and then normalized.

Using this wave function for a given triplet of $\alpha = \alpha_k$, $\beta = \beta_k$, and $\gamma = \gamma_k$ in the sum in Equation (9), the matrix element for ionizing either positron under the influence of the length dipole operator $(z_1 + z_2)$ is the sum of four terms:

$$\begin{aligned} \mu_{PI} = 2\,(I_{11} + I_{21})\,\frac{\lambda^{3/2}}{\sqrt{2\pi}} &= \int\!\!\int d^3x_1 d^3x_2 e^{-\alpha r_1 - \beta r_2 - \gamma r_{12}}\,(z_1 + z_2)\,\frac{\lambda^{3/2}}{\sqrt{\pi}}\frac{1}{\sqrt{2}}, \\ &\times \left[e^{-\lambda r_2} e^{ikz_1} + e^{-\lambda r_1} e^{ikz_2}\right] \end{aligned} \tag{11}$$

where the first factor of two comes from $I_{11} = I_{22}$ and so on, whose subscripts j refer to z_j in the dipole operator and in the plane wave, respectively, and we have factored out the coefficient $\frac{\lambda^{3/2}}{\sqrt{2\pi}}$ that is common to all terms. We keep λ, the magnitude of the charge of the hydrogen nucleus, in symbolic form rather than setting it to one so that we can take derivatives of it to represent powers of r_j.

The cross section for radiatively attaching a second positron to $\overline{\mathrm{H}}\,(1s)$ to create the $(1s^2\,{}^1S^e)$ state of the $\overline{\mathrm{H}}^+$ ion, via the reaction in Equation (1), can be obtained from the principle of detailed balance (see, e.g., Landau and Lifshitz [55]) following the lead of Drake [56] and then Jacobs, Bhatia, and Temkin [57], who applied the principle of detailed balance to obtain the radiative attachment coefficient (for an electron) to form the $(2p^2\,{}^3P^e)$ metastable H^- state from H $(2s, 2p)$. For the $(1s^2\,{}^1S^e)$ case, we have [16],

$$\sigma_{RA}(k) = \frac{g_1 p_\omega^2}{g_2 p_e^2}\sigma_{PI} = \frac{6\alpha^2\left(k^2+\gamma_0^2\right)^2}{12\cdot 2^2 k^2}\sigma_{PI}\,, \tag{12}$$

where $g_1/g_2 = 6/12$ is the statistical weight ratio. Here, the photon momentum relative to the ion is given by $p_\omega = \hbar\omega/c = (k^2+\gamma_0^2)/2c$, and p_e is the positron momentum k. Note that c in atomic units is the inverse of the fine structure constant α.

To estimate formation rates of $\overline{\mathrm{H}}^+$, it is helpful to calculate the positron attachment to $\overline{\mathrm{H}}$ as a function of temperature rather than energy, as is common in astrophysical applications. This rate coefficient α_{RA} is formed as the expectation value of $v\sigma_{RA}$ with the normalized Maxwell–Boltzmann distribution $f(v)$ as,

$$\begin{aligned} \alpha_{RA}(T) = \langle v\sigma_{RA}\rangle &= 4\pi\int_0^\infty dv\, v\sigma_{RA}\left(k\left(v\right)\right)\left(\frac{m}{2\pi k_B T}\right)^{3/2} v^2 \exp\left[-mv^2/\left(2k_B T\right)\right] \\ &= \sqrt{\frac{8k_B T}{m_e}}\frac{1}{\sqrt{\pi}}\int_0^\infty dx\, x\frac{g_{PI}}{g_{RA}}\frac{P_\omega^2}{p_e^2}\sigma_{PI}\left(\sqrt{2k_B T x}\right)\exp\left[-x\right]\,. \end{aligned} \tag{13}$$

whose overall coefficient is

$$\begin{aligned} \sqrt{\frac{8k_B T}{m_e}} &= \sqrt{\frac{8\times 8.61733262\times 10^{-5}\ \mathrm{eV}\ K^{-1}T}{0.51099895000\times 10^6\ \mathrm{eV}}}299792458\times 10^2\ \mathrm{cm/s} \\ &= 1.10113894\times 10^6\ \mathrm{cm/s}\sqrt{K^{-1}T} \end{aligned} \tag{14}$$

where the temperature T is given in Kelvins K.

3. Evaluating Integrals

The algebra in each case is greatly reduced by making the replacement

$$e^{-\gamma r_{12}}e^{-r_2(\beta+\lambda)} = \left(-\frac{\partial}{\partial\gamma}\right)\left(-\frac{\partial}{\partial\lambda}\right)\frac{e^{-r_2(\beta+\lambda)}}{r_2}\frac{e^{-\gamma r_{12}}}{r_{12}} \tag{15}$$

in each term, with the notation

$$I_{j1} = \left(\frac{\partial}{\partial\gamma}\right)\left(\frac{\partial}{\partial\lambda}\right)R_{0j1} \tag{16}$$

where

$$\begin{aligned} R_{011} &= \int\int d^3x_1 d^3x_2 e^{-\alpha r_1}\frac{e^{-r_2(\beta+\lambda)}}{r_2}\frac{e^{-\gamma r_{12}}}{r_{12}}\left(z_1 e^{ikz_1}\right) \\ R_{021} &= \int\int d^3x_1 d^3x_2 e^{-\alpha r_1}\frac{e^{-r_2(\beta+\lambda)}}{r_2}\frac{e^{-\gamma r_{12}}}{r_{12}}\left(z_2 e^{ikz_1}\right)\,. \end{aligned} \tag{17}$$

For the cross-terms $\left(z_2 e^{ikz_1}\right) = \left(r_2 \cos\theta_2 e^{ikr_1 \cos\theta_1}\right)$, we need the conversion given in Ley-Koo and Bunge [58]

$$Y_{k,m}(\theta_2, \phi_2) = \sum_{M=-k}^{k} \left[\mathfrak{D}^{(k)}_{m,M}(0, \theta_1, \phi_1)\right]^* Y_{k,M}(\theta_{12}, \phi_{12}) \tag{18}$$

where from Edmonds [59] we have

$$\begin{aligned} Y_{1,0}(\theta_2, \phi_2) &= \sum_{M=-1}^{1} \left[\mathfrak{D}^{(1)}_{0,M}(0, \theta_1, \phi_1)\right]^* Y_{1,M}(\theta_{12}, \phi 1_2) \quad , \\ &= \sum_{M=-1}^{1} \left[\sqrt{\frac{4\pi}{3}} Y_{1,M}(\theta_1, \phi_1)\right]^* Y_{1,M}(\theta_{12}, \phi_{12}) \end{aligned} \tag{19}$$

so that in this case

$$\begin{aligned} \cos\theta_2 = 2\sqrt{\frac{\pi}{3}} Y_{1,0}(\theta_2, \phi_2) &= \tfrac{4\pi}{3} \textstyle\sum_{M=-1}^{1} Y^*_{1,M}(\theta_1, \phi_1)\, Y_{1,M}(\theta_{12}, \phi_{12}) \\ = P_1(\cos\theta_2) &= \cos\theta_1 \cos\theta_{12} + \sin\theta_1 \sin\theta_{12} \cos(\phi_1 + \phi_{12}) \end{aligned} \tag{20}$$

we recover the law of cosines. Ley-Koo and Bunge [58] note that the only contribution comes from the term with $M = 0$, "because the point of interest is on the polar axis" so that only the cosine-product term remains, which we confirmed by calculating both terms.

Introducing an addition theorem for the plane wave ([60], p. 671, Equation (B.44)) helps us to do the first angular integral in each of the terms in Equation (17),

$$\int d\Omega_1 \left(r_1 P_1(\cos\theta_1)\right) \left(\sum_{l=0}^{\infty} (2l+1)\, i^l j_l(kr_1) P_l(\cos\theta_1)\right) = \left(r_1 \frac{2}{(2+1)} 2\pi\right) \left((2+1)\, i^1 j_1(kr_1)\right) \quad . \tag{21}$$

All expressions that follow should in principle be multiplied by this factor of i, but, since we take the absolute square of sums of these transition amplitudes to get the cross section, we ignore this factor.

We follow Ley-Koo and Bunge [58] in replacing $d\Omega_2$—the differential solid angle around $\hat{r}_2$ in a frame of reference in which $\mathbf{r}_1$ is taken as the polar axis—by $d\Omega_{12} = \sin\theta_{12} d\theta_{12} d\phi_{12}$. One can immediately integrate over $d\phi_{12}$. They change variables to $\cos\theta_{12} = \left(r_1^2 + r_2^2 - r_{12}^2\right) / (2r_1 r_2)$ giving $\sin\theta_{12} d\theta_{12} = (-2r_{12})\, dr_{12} / (2r_1 r_2)$, but one may also change variables to $\cos\theta_{12} = u_{12}$ giving $\sin\theta_{12} d\theta_{12} = du_{12}$ and simply do that integral using integrals we have not found in the literature:

$$\begin{aligned} \int_{-1}^{1} du_{12} \frac{e^{-\gamma\sqrt{r_1^2 - 2r_1 r_2 u_{12} + r_2^2}}}{\sqrt{r_1^2 - 2r_1 r_2 u_{12} + r_2^2}} &= \frac{e^{-\gamma|r_1 - r_2|} - e^{-\gamma|r_1 + r_2|}}{\gamma r_1 r_2} \\ \int_{-1}^{1} du_{12} \frac{e^{-\gamma\sqrt{r_1^2 - 2r_1 r_2 u_{12} + r_2^2}}}{\sqrt{r_1^2 - 2r_1 r_2 u_{12} + r_2^2}} u_{12} &= \frac{1}{\gamma^3 r_1^2 r_2^2} \left(\left(e^{-\gamma|r_1 - r_2|} + e^{-\gamma(r_1 + r_2)}\right) r_1 r_2 \gamma^2 - e^{-\gamma|r_1 - r_2|} + e^{-\gamma r_1 - \gamma r_2}\right. \\ &+ \left.\gamma\left(e^{-\gamma(r_1 + r_2)}(r_1 + r_2) - e^{-\gamma|r_1 - r_2|}|r_1 - r_2|\right)\right) \quad [r_1 > 0,\ r_2 > 0], \\ \frac{1}{\gamma r_1 r_2} \frac{\partial}{\partial a} \int_{-1}^{1} du_{12}\, e^{-\gamma\sqrt{r_1^2 - a2r_1 r_2 u_{12} + r_2^2}}\bigg|_{a=1} u_{12} &= \frac{1}{\gamma r_1 r_2} \frac{\partial}{\partial a} \frac{1}{a^2 \gamma^4 r_1^2 r_2^2} \left(e^{-\gamma\sqrt{r_1^2 - 2ar_2 r_1 + r_2^2}} \left(-\left(r_1^2 - 3ar_2 r_1 + r_2^2\right)\gamma^2\right.\right. \\ &+ \left|\left(a\gamma^2 r_1 r_2 - 3\right)\sqrt{r_1^2 - 2ar_2 r_1 + r_2^2}\gamma - 3\right) + e^{-\gamma\sqrt{r_1^2 + 2ar_2 r_1 + r_2^2}} \\ &\times \left.\left(\left(r_1^2 + 3ar_2 r_1 + r_2^2\right)\gamma^2 + \left(ar_1 r_2 \gamma^2 + 3\right)\sqrt{r_1^2 + 2ar_2 r_1 + r_2^2}\gamma + 3\right)\right)_{a=1} \end{aligned} \tag{22}$$

The last integrals to do for the cross section are

$$
\begin{aligned}
R_{011} &= 8\pi^2 \frac{2}{(\beta+\lambda)^2-\gamma^2}\int_0^\infty dr_1 r_1^2 j_1(kr_1)\left(e^{-(\alpha+\beta+\lambda)r_1}-e^{-(\alpha+\gamma)r_1}\right) \quad . \qquad (23)\\
R_{021} &= 8\pi^2 \frac{4}{((\beta+\lambda)^2-\gamma^2)^2}\int_0^\infty dr_1\left(j_1(kr_1)\left(e^{-(\alpha+\beta+\lambda)r_1}-e^{-(\alpha+\gamma)r_1}\right)\right)\\
&+ 8\pi^2 \frac{4}{((\beta+\lambda)^2-\gamma^2)^2}\int_0^\infty dr_1\left(r_1 j_1(kr_1)\left((\beta+\lambda)e^{-(\alpha+\beta+\lambda)r_1}-\gamma e^{-(\alpha+\gamma)r_1}\right)\right)\\
&+ 8\pi^2 \frac{2}{(\beta+\lambda)^2-\gamma^2}\int_0^\infty dr_1\left(r_1^2 j_1(kr_1)\, e^{-(\alpha+\beta+\lambda)r_1}\right)
\end{aligned}
$$

These are easily done and the derivatives in Equation (16) taken, providing the core of the numerical Maxwell–Boltzmann integral in Equation (13) for the rate coefficient α_{RA} for producing the negative hydrogen ion H^- or its antimatter equivalent, the positive antihydrogen ion $\overline{H}^+$ [16].

4. Doing the r_1 Integrals Last

The conventional path to crafting an analytical rate coefficient α_{RA} for producing the positive antihydrogen ion $\overline{H}^+$, or its matter equivalent, would be to integrate pair-products of terms in the analytical results for the cross section found from Equation (23) above, or as given in Keating's master's thesis [21], reproduced in Equation (2). This latter form organizes the sums of terms in the cross section most compactly, but integrating every pair-product of every term in even this set (Equation (2)) would require some two-dozen analytical integrals such as

$$
\int_0^\infty \frac{e^{-x}\left(\gamma_0{}^2+2k_BTx\right)^3}{4\sqrt{2}(k_BT)^{5/2}x^{3/2}}\tan^{-1}\left(\frac{\sqrt{2k_BTx}}{p_f}\right)\tan^{-1}\left(\frac{\sqrt{2k_BTx}}{p_i}\right)dx \qquad (24)
$$

and

$$
\int_0^\infty \frac{e^{-x}\left(\gamma_0{}^2+2k_BTx\right)^3\tan^{-1}\left(\frac{\sqrt{2k_BTx}}{p_i}\right)}{4k_B^2T^2x\left(p_f^2+2k_BTx\right)}dx \quad , \qquad (25)
$$

few of which are easily done.

Instead of this obvious approach, we take the road less traveled and take these integrals in reverse order because of the uniformity of the integrands in Equation (23). The downside of this novel approach is that we must form the product of distinct radial integrals rather than squaring the analytical result of the result of a single integration, and there are many dead ends on a path to integrating over both of these radial variables after integrating over the equivalent of x in Equation (13). We did, however, find a means to do so, as follows.

We first take the derivatives in Equation (16) of Equation (23) to obtain the requisite terms of Equation (11), after substituting the conventional Bessel function for the spherical Bessel function $\frac{1}{\sqrt{kr_1}}\sqrt{\frac{\pi}{2}}J_{\frac{3}{2}}(kr_1)=j_1(kr_1)$ ([60], p. 673, Equation (C.2)):

$$
\begin{aligned}
I_{11} &= 4\pi\int_0^\infty dr_1 4\pi\left(\frac{1}{\sqrt{k}}\sqrt{\frac{\pi}{2}}J_{\frac{3}{2}}(kr_1)\, r_1^{3/2}\right)\\
&\times\left(\frac{8\gamma(\beta+\lambda)e^{-(\alpha+\beta+\lambda)r_1}}{((\beta+\lambda)^2-\gamma^2)^3}-\frac{8\gamma(\beta+\lambda)e^{-(\alpha+\gamma)r_1}}{((\beta+\lambda)^2-\gamma^2)^3}+\frac{2r_1\gamma e^{-(\alpha+\beta+\lambda)r_1}}{((\beta+\lambda)^2-\gamma^2)^2}+\frac{2r_1(\beta+\lambda)e^{-(\alpha+\gamma)r_1}}{((\beta+\lambda)^2-\gamma^2)^2}\right) \qquad (26)
\end{aligned}
$$

and

$$
\begin{aligned}
I_{21} &= 4\pi\int_0^\infty dr_1 4\pi\left(\frac{1}{\sqrt{k}}\sqrt{\frac{\pi}{2}}J_{\frac{3}{2}}(kr_1)\, r_1^{3/2}\right)\left(\frac{48\gamma(\beta+\lambda)e^{-(\alpha+\beta+\lambda)r_1}}{r_1^2((\beta+\lambda)^2-\gamma^2)^4}-\frac{48\gamma e^{-(\alpha+\gamma)r_1}(\beta+\lambda)}{r_1^2((\beta+\lambda)^2-\gamma^2)^4}+\frac{48\gamma(\beta+\lambda)^2e^{-(\alpha+\beta+\lambda)r_1}}{r_1((\beta+\lambda)^2-\gamma^2)^4}\right.\\
&\left.-\frac{48\gamma^2(\beta+\lambda)e^{-(\alpha+\gamma)r_1}}{r_1((\beta+\lambda)^2-\gamma^2)^4}+\frac{16\gamma(\beta+\lambda)e^{-(\alpha+\beta+\lambda)r_1}}{((\beta+\lambda)^2-\gamma^2)^3}+\frac{8\gamma(\beta+\lambda)e^{-(\alpha+\gamma)r_1}}{((\beta+\lambda)^2-\gamma^2)^3}+\frac{2r_1\gamma e^{-(\alpha+\beta+\lambda)r_1}}{((\beta+\lambda)^2-\gamma^2)^2}\right) \quad . \qquad (27)
\end{aligned}
$$

The first term of I_{11} may be combined with third-to-last term of I_{21}, as may the third term of I_{11}, with the last term of I_{21}. The second term of I_{11} and the second-to-last of I_{21} cancel, thus one is left with seven terms comprising three powers of r_1 with two kinds of exponentials, a considerably uniform set of analytic functions to be integrated.

If one wishes to do the Maxwell–Boltzmann integral of products of such functions first, the products must be written as unique integrals. That is, the rate coefficient will be sums of terms

$$
\begin{aligned}
B(\mathrm{T}, \mathrm{a}, \sigma, \mathrm{h}, \tau, \mathrm{j}, \beta, \gamma, \lambda, \alpha, \gamma_b, \mathrm{ps}, \mathrm{pt}) \quad &= \frac{6.811556\times 10^{-20}\left(1.10114\times 10^{6}\sqrt{T}\right)}{\sqrt{\pi}\,(2^2c^2)}\frac{g_{PI}}{g_{RA}} \\
&\times 128\pi^4\lambda^3 \int_0^\infty dR_1 \int_0^\infty dr_1 \int_0^\infty dx \frac{r_1^{h+\frac{3}{2}} R_1^{j+\frac{3}{2}} \left(2k_BTx+\gamma_0^2\right)^3 J_{\frac{3}{2}}\left(\sqrt{2}\sqrt{k_BTx}r_1\right) J_{\frac{3}{2}}\left(\sqrt{2}\sqrt{k_BTx}R_1\right) e^{-\sigma r_1-\tau R_1-ax}}{k_BT\left((d+\lambda)^2-\gamma^2\right)^{\mathrm{ps}}\left((f+\lambda)^2-\gamma_b^2\right)^{\mathrm{pt}}}
\end{aligned}
\tag{28}
$$

where we replace β in the denominators with d and f to allow for multiplication of sums of terms whose positrons are exchanged by the $\hat{P}_{12}$ permutation operator in Equation (9) for the two identical positrons $\alpha \leftrightarrow \beta$, as well as different values of α and β arising from the various terms in the adjoining sum over terms in the wave function. The latter also explains the need to distinguish γ_b from γ in the second denominator. It turned out that the a in the exponential was unneeded for the present problem, but we have left it in for those who might need this sort of integral for another problem and later set it to equal one.

One may perform the x and then the r_1 integrals as they stand, but the resulting expression did not allow for integration over R_1. This is, of course, why one normally would never do the integrals in this order if it could be avoided. However, the hope of an overall simpler summing of products of terms if this unusual and difficult integration order is successful eventually found fruit in a series approach. We first express the Bessel functions in terms of the hypergeometric function [61–63]

$$
J_{\frac{3}{2}}(kr_1) = \frac{1}{\Gamma\left(\frac{5}{2}\right)}\left(\frac{kr_1}{2}\right)^{3/2} {}_0F_1\left(;\frac{5}{2};-\frac{1}{4}(kr_1)^2\right) \tag{29}
$$

and combine their product as [64,65]

$$
\begin{aligned}
&\frac{1}{\Gamma\left(\frac{5}{2}\right)}\left(\frac{r_1\sqrt{2k_BTx}}{2}\right)^{3/2} {}_0F_1\left(;\tfrac{5}{2};-\tfrac{1}{4}\left(r_1\sqrt{2k_BTx}\right)^2\right)\frac{1}{\Gamma\left(\frac{5}{2}\right)}\left(\frac{R_1\sqrt{2k_BTx}}{2}\right)^{3/2} {}_0F_1\left(;\tfrac{5}{2};-\tfrac{1}{4}\left(R_1\sqrt{2k_BTx}\right)^2\right) \\
= \;&\frac{4\sqrt{2}\left(r_1\sqrt{k_BTx}\right)^{3/2}\left(R_1\sqrt{k_BTx}\right)^{3/2}}{9\pi}\sum_{m=0}^{\infty}\frac{2^{-m}\left(-r_1^2\right)^m(k_BTx)^m}{m!\left(\frac{5}{2}\right)_m} {}_2F_1\left(-m-\tfrac{3}{2},-m;\tfrac{5}{2};\tfrac{R_1^2}{r_1^2}\right) \quad .
\end{aligned}
\tag{30}
$$

After expanding

$$
\left(2k_BTx+\gamma_0^2\right)^3 = 8k_B^3T^3x^3+12k_B^2T^2x^2\gamma 0^2+6k_BTx\gamma 0^4+\gamma 0^6 \tag{31}
$$

the integral over powers n of x in this sum is [66] (p. 364 No. 3.381.4)

$$
\int_0^\infty dx e^{-ax}x^{m+n+\nu} = a^{-m-n-\nu-1}\Gamma(m+n+\nu+1)
$$

The r_1 integral, including coefficients from the second line of Equation (28), for each term in the m-sum follows as [67]

$$\begin{aligned}
B_2 &= \pi^3 2^{\frac{19}{2}-m} \lambda^3 \frac{a^{-m-n-\frac{5}{2}} (k_B T)^{m+\frac{3}{2}} \Gamma\left(m+n+\frac{5}{2}\right)}{9\mathrm{kB}Tm! \left(\frac{5}{2}\right)_m ((d+\lambda)^2-\gamma^2)\left((f+\lambda)^2-\gamma b^2\right)} R_1^{j+3} e^{-\tau R_1} \int_0^\infty dr_1 r_1^{h+3} \left(-r_1^2\right)^m e^{-\sigma r_1} {}_2F_1\left(-m-\frac{3}{2}, -m; \frac{5}{2}; \frac{R_1^2}{r_1^2}\right) \\
&= \pi^{7/2} (-1)^m 2^{m+\frac{25}{2}} \lambda^3 R_1^{j+3} e^{-\tau R_1} a^{-m-n-\frac{5}{2}} \sigma^{-h-2m-4} (k_B T)^{m+\frac{1}{2}} \Gamma\left(m+n+\frac{5}{2}\right) \\
&\times \frac{1}{9m! \left(\frac{5}{2}\right)_m \Gamma\left(-m-\frac{3}{2}\right) ((d+\lambda)^2-\gamma^2)\left((f+\lambda)^2-\gamma b^2\right)} \\
&\times \left(-2(m+1) \frac{\Gamma(-2m-3)}{\Gamma(-m)} \Gamma(h+2m+4) {}_2F_3\left(-m-\frac{3}{2}, -m; \frac{5}{2}, -\frac{h}{2}-m-\frac{3}{2}, -\frac{h}{2}-m-1; \frac{1}{4}\sigma^2 R_1^2\right)\right. \\
&+ 3R_1^4 \left(-\frac{1}{R_1^2}\right)^{-\frac{h}{2}-m} \sigma^{h+2m+4} \left[-\frac{(h+2)\Gamma(h+1)\cos\left(\frac{\pi h}{2}+\pi m\right)}{h+2m+7} \frac{\Gamma(-h-2m-5)}{\Gamma(-m)} {}_2F_3\left(\frac{h}{2}+\frac{1}{2}, \frac{h}{2}+2; \frac{1}{2}, \frac{h}{2}+m+3, \frac{h}{2}+m+\frac{9}{2}; \frac{1}{4}\sigma^2 R_1^2\right)\right. \\
&- \left.\left.\frac{(h+3)\sigma\Gamma(h+2)\sin\left(\frac{\pi h}{2}+\pi m\right)}{\sqrt{-\frac{1}{R_1^2}}(h+2m+8)} \frac{\Gamma(-h-2m-6)}{\Gamma(-m)} {}_2F_3\left(\frac{h}{2}+1, \frac{h}{2}+\frac{5}{2}; \frac{3}{2}, \frac{h}{2}+m+\frac{7}{2}, \frac{h}{2}+m+5; \frac{1}{4}\sigma^2 R_1^2\right)\right]\right. . \\
& [\Re(h)+1>0 \wedge \Re(h+2m)+4>0 \wedge \Re(\sigma)>0 \wedge (R_1^2 \notin \mathbb{R} \vee \Re(R_1^2) \leq 0) \wedge (R_1 \notin \mathbb{R} \vee (\Re(R_1)=0 \wedge R_1 \neq 0))]
\end{aligned} \tag{32}$$

We ignored the restriction against R_1 being an element of the reals under the assumption that this result could likely be considered as a distribution whose integral would smooth out any singularities arising from this restriction. That assumption paid off. Let us consider the gamma functions of negative integer arguments that append each of the ${}_2F_3$ functions ([66], p. 946 No. 8.334.3; [68]),

$$\begin{aligned}
\Gamma(-2m-3) &= -\frac{\pi \csc(\pi(2m+3))}{\Gamma(2m+4)} \\
\Gamma(-m) &= -\frac{\pi \csc(\pi m)}{\Gamma(m+1)} \\
\Gamma(-h-2m-5) &= -\frac{\pi \csc(\pi(h+2m+5))}{\Gamma(h+2m+6)} \\
\Gamma(-h-2m-6) &= -\frac{\pi \csc(\pi(h+2m+6))}{\Gamma(h+2m+7)}
\end{aligned} \tag{33}$$

whose ratios have the following values for m an integer:

$$\begin{aligned}
-\frac{\pi \csc(\pi(2m+3))}{\Gamma(2m+4)} \left(-\frac{\pi \csc(\pi m)}{\Gamma(m+1)}\right)^{-1} &= \frac{(-1)^{m+1}\Gamma(m+1)}{2\Gamma(2m+4)} \\
-\frac{\pi \csc(\pi(h+2m+5))}{\Gamma(h+2m+6)} \left(-\frac{\pi \csc(\pi m)}{\Gamma(m+1)}\right)^{-1} &= 0 \\
-\frac{\pi \csc(\pi(h+2m+6))}{\Gamma(h+2m+7)} \left(-\frac{\pi \csc(\pi m)}{\Gamma(m+1)}\right)^{-1} &= 0
\end{aligned}$$

Integrating the one remaining term (under the restriction $\Re(j) > -4 \wedge ((\Re(h+j) < -2 \wedge \Re(\sigma-\tau) \leq 0 \wedge \Re(\sigma+\tau) \geq 0) \vee (\Re(h+j) \geq -2 \wedge \Re(\sigma-\tau) < 0 \wedge \Re(\sigma+\tau) > 0)) \wedge (\Re(\tau) > 0 \vee (\Re(\tau) = 0 \wedge \Re(j+2m) < -6))$ gives us our final result,

$$
\begin{aligned}
B(\mathrm{T}, \mathrm{a}, \sigma, \mathrm{h}, \tau, \mathrm{j}, \beta, \gamma, \lambda, \alpha, \gamma_b, \mathrm{ps}, \mathrm{pt}) = \; & \frac{6.811556\times 10^{-20}\left(1.10114 \times 10^6 \sqrt{T}\right)}{\sqrt{\pi}\,(2^2 c^2)} \frac{g_{PI}}{g_{RA}} \\
\times \; & (k_B T)^{\frac{1}{2}} \frac{\pi^{7/2}\lambda^3}{((d+\lambda)^2-\gamma^2)^{\mathrm{ps}}\,((f+\lambda)^2-\gamma_b^2)^{\mathrm{pt}}} \\
\times \; & \sum_{m=0}^{\infty} \frac{\left(2^{m+25/2}(m+1)a^{-m-\frac{5}{2}}\Gamma(m+1)(k_B T)^m\right)}{9m!\left(\frac{5}{2}\right)_m \Gamma\left(-m-\frac{3}{2}\right)\Gamma(2m+4)} \\
\times \; & \left(\frac{8k_B^3 T^3 \Gamma\left(m+\frac{5}{2}+3\right)}{a^3} + \frac{12k_B^2 T^2 \gamma_0^2 \Gamma\left(m+\frac{5}{2}+2\right)}{a^2}\right. \\
+ \; & \left.\frac{6k_B T \gamma_0^4 \Gamma\left(m+\frac{5}{2}+1\right)}{a} + a^0 \gamma_0^6 \Gamma\left(m+\frac{5}{2}+0\right)\right) \\
\times \; & \tau^{-j-4}\Gamma(j+4)\sigma^{-h-2m-4}\Gamma(h+2m+4) \\
\times \; & {}_4F_3\left(\frac{j}{2}+2, \frac{j}{2}+\frac{5}{2}, -m-\frac{3}{2}, -m; \frac{5}{2}, -\frac{h}{2}-m-\frac{3}{2}, -\frac{h}{2}-m-1; \frac{\sigma^2}{\tau^2}\right)
\end{aligned}
\quad (34)
$$

The rate coefficient is then a quadruple sum. First, we have a sum over the seven terms in $I_{11} + I_{21}$ of Equations (26) and (27) plus their seven permutations $\alpha \leftrightarrow \beta$, all taking on appropriate values of σ, h, and *ps* as used in Equation (28), and being multiplied by the appropriate numerators in Equations (26) and (27) such as $48\gamma(\beta+\lambda)$. Second, we have a sum over the corresponding 14 terms of the second factor that takes on the values of τ, j, and pt, multiplied by the corresponding numerators. Third, we have a sum over the n terms in the wave function having differing values for the parameters α, β, and γ, and that term's coefficient c whose value is found by diagonalizing the Hamiltonian matrix in order to minimize the ground state energy and then normalized. We display the first and last elements of the fifty-term version we used for both analytical and numerical calculations in case readers wish to check it against their own derivations:

$$
\begin{aligned}
\alpha(1) &= 0.6878357596671101, \cdots, & \alpha(50) &= 0.37080904876118337 \\
\beta(1) &= 1.2354854281591452, \cdots, & \beta(50) &= 1.1073078257847453 \\
\gamma(1) &= 0.012984468708341133, \cdots, & \gamma(50) &= 0.28320160279252 \\
\mathrm{c}(1) &= -2.534888772248287, \cdots, & c(50) &= 0.02873436866901307
\end{aligned}
\quad . \quad (35)
$$

The final sum is again over this same wave-function set, but this time associated with the second factor τ. Because the off-diagonal terms of these four sums are symmetrical, it is more efficient computationally to account for that.

5. Comparison with Numerical Integration

Table 1 shows the comparison of the present analytical rate coefficient α_{RA} for attaching a positron to antihydrogen to form $\overline{\mathrm{H}}^+$ to our prior rate coefficient found via numerical integration [16], at temperatures ranging from 1 to 400 K. There we found that, for a fully two-positron 200-term wave function [18] composed of explicitly correlated exponentials and for T $\lesssim$ 6 K, the rate coefficient is essentially linear and may be fit by $\alpha_{RA} = 0.001050 \times 10^{-15}$ cm^{3}s^{-1} T K^{-1}. One sees the potential for this linear behavior in the analytic result in Equation (34) in the factor outside of the sum, but only if the $m=0$ term is the dominant contributor to the sum at this temperature. For a fifty-term wave function, the first three terms in the analytical sum are $\left(0.00106039 - 1.2144278 \times 10^{-6} + 1.203002 \times 10^{-9}\right) \times 10^{-15}$ cm^{3}s^{-1} $T\ K^{-1}$, which sum to a value $0.00105917 \times 10^{-15}$ cm^{3}s^{-1} $T\ K^{-1}$ and the numerical integral for this same fifty-term wave function gives $0.00102625 \times 10^{-15}$ cm^{3}s^{-1} $T\ K^{-1}$, a 3% difference.

As the temperature increases, this series result that involves powers of the temperature will eventually fail. For a $Z = 1$ ion, we find that there is no convergence of the m series at T=700 K. For higher Z, higher temperatures are likely possible.

Table 1. The rate coefficient α_{RA} (10^{-15} cm^{3}s^{-1}) for attaching a positron to antihydrogen to form $\overline{\mathrm{H}}^+$ for a fifty-term wave function.

Positron Temperature (K)	Analytical Integration	Number of Terms in m-Sum (34)	Numerical Integration
1	0.001059	3	0.001026
10	0.01056	4	0.01053
100	0.1030	15	0.1030
400	0.3809	15	0.3810

6. Discussion and Concluding Remarks

We found the analytic form for the rate coefficient α_{RA} for attaching a positron to antihydrogen to form $\overline{\mathrm{H}}^+$ as a series convergent for temperatures of 400 K and below, which may be used to estimate formation rates. As our result is for attachment from the 1s antihydrogen state, it depends on the trapped antihydrogen having been held for long enough to ensure that it will have decayed to the ground state from the likely excited state in which it is formed [7,8].

If the positron plasma is held in a Penning trap of the type used to form antihydrogen, such as those reviewed in [69] at a density of $n_e = 10^{16}$ m^{-3} in a magnetic field of 1 T at a sub-mm plasma radius, then the positron speeds due to rotation of the plasma can be neglected. The temperature range currently used to form and trap antihydrogen is in the range of 10s of K or lower. At 100 K, the rate coefficient is $\alpha_{RA} = 10 \times 10^{-17}$ cm^{3}s^{-1}, and if there is unit overlap between the positron plasma and the antihydrogen, then the reaction rate, the product $n_e\, \alpha_{RA}$, at this temperature would be 10×10^{-7} s^{-1} per antihydrogen atom. As the temperature falls to 10 K, the reaction rate falls in an essentially linear manner to be 1×10^{-7} s^{-1} per antihydrogen atom. This might just be observable, given the long antihydrogen storage times achieved by ALPHA [8], if all ALPHA's antiprotons could be converted into trapped $\overline{\mathrm{H}}$s, while still allowing the anti-atoms to interact with warm positron clouds. Cold $\overline{p}$ numbers, and hopefully those of trapped $\overline{\mathrm{H}}$ as well, will likely be enhanced by around a factor of 10^2 within the next three years as CERN's AD facility is enhanced by the addition of a further storage ring, ELENA (see, e.g., [70]) that will deliver antiprotons to experiments at an energy nearer 100 keV than the current 5 MeV.

Funding: This research received no external funding.

Conflicts of Interest: The author declares that there is no conflict of interest.

References and Note

1. Maury, S. The Antiproton Decelerator: AD. *Hyperfine Interactions* **1997**, *109*, 43–52. [CrossRef]
2. Amoretti, M.; et al. (ATHENA Collaboration). Production and detection of cold antihydrogen atoms. *Nature* **2002**, *419*, 456–459. [CrossRef] [PubMed]
3. Gabrielse, G.; et al. (ATRAP Collaboration). Background-Free Observation of Cold Antihydrogen with Field-Ionization Analysis of Its States. *Phys. Rev. Lett.* **2002**, *89*, 213401. [CrossRef]
4. Enomoto, Y.; Kuroda, N.; Michishio, K.; Kim, C.H.; Higaki, H.; Nagata, Y.; Kanai, Y.; Torii, H.A.; Corradini, M.; Leali, N.; et al. Synthesis of Cold Antihydrogen in a Cusp Trap. *Phys. Rev. Lett.* **2010**, *105*, 243401. [CrossRef] [PubMed]
5. Andresen, G.B.; et al. (ALPHA Collaboration). Trapped antihydrogen. *Nature* **2010**, *468*, 673–676. [CrossRef]
6. Andresen, G.B.; et al. (ALPHA Collaboration). Search for trapped antihydrogen. *Phys. Lett. B* **2011**, *695*, 95–104. [CrossRef]

7. Gabrielse, G.; et al. (ATRAP Collaboration). Trapped Antihydrogen in Its Ground State. *Phys. Rev. Lett.* **2012**, *108*, 113002. [CrossRef]
8. Andresen, G.B.; et al. (ALPHA Collaboration). Confinement of antihydrogen for 1000 seconds. *Nat. Phys.* **2011**, *7*, 558–564.
9. Amole, C.; et al. (ALPHA Collaboration). Resonant quantum transitions in trapped antihydrogen atoms. *Nature* **2012**, *483*, 439–443. [CrossRef]
10. Amole, C.; et al. (ALPHA Collaboration). An experimental limit on the charge of antihydrogen. *Nat. Commun.* **2014**, *5*, 3955. [CrossRef]
11. The ALPHA Collaboration; Charman, A.E. Description and first application of a new technique to measure the gravitational mass of antihydrogen. *Nat. Commun.* **2013**, *4*, 1875. [CrossRef] [PubMed]
12. Walz, J.; Hansch, T.W. A Proposal to Measure Antimatter Gravity Using Ultracold Antihydrogen Atoms. *Gen. Rel. Grav.* **2004**, *150*, 561–570. [CrossRef]
13. Perez, P.; Sacquin, Y. The GBAR experiment: Gravitational behaviour of antihydrogen at rest. *Class. Quant. Grav.* **2012**, *87*, 184008. [CrossRef]
14. Van der Werf, D.P. The GBAR experiment. *Int. J. Mod. Phys. Conf. Ser.* **2014**, *30*, 1460263. [CrossRef]
15. Keating, C.M.; Charlton, M.; Straton, J.C. On the production of the positive antihydrogen ion $\overline{H}^+$ via radiative attachment. *J. Phys. B At. Mol. Opt. Phys.* **2014**, *47*, 225202. [CrossRef]
16. Keating, C.M.; Pak, K.Y.; Straton, J.C. Producing the positive antihydrogen ion $\overline{H}^+$ via radiative attachment. *J. Phys. B At. Mol. Opt. Phys.* **2016**, *49*, 074002. [CrossRef]
17. Ohmura, T.; Ohmura, H. Electron-Hydrogen Scattering at Low Energies. *Phys. Rev.* **1960**, *118*, 154. [CrossRef]
18. Cann, N.M. An Accurate Treatment of Heliumlike Ions: Properties and Wavefu net ions. Ph.D. Thesis, Dalhousie University, Halifax, NS, Canada, 1993.
19. Thakkar, A.J.; Smith, V.H. Compact and accurate integral-transform wave functions. I. The 1^1S state of the helium-like ions from H^- through Mg^{10+}. *Phys. Rev. A* **1977**, *15*, 1–15. [CrossRef]
20. Bhatia, A.K. Hybrid theory of P-wave electron-Li^{2+} elastic scattering and photoabsorption in two-electron systems. *Phys. Rev. A* **2013**, *87*, 042705. [CrossRef]
21. Keating, C.M. A Method for Achieving Analytic Formulas for Three Body Integrals Consisting of Powers and Exponentials in All Three Interparticle Hyllerass Coordinates. Master's Thesis, Portland State University, Portland, OR, USA, 2015.
22. Wildt, R. Electron Affinity in Astrophysics. *Astrophys. J.* **1938**, *89*, 295–301. [CrossRef]
23. Wildt, R. Negative Ions of Hydrogen and the Opacity of Stellar Atmospheres. *Astrophys. J.* **1939**, *90*, 611. [CrossRef]
24. Chandrasekhar, S. On the Continuous Absorption Coefficient of the Negative Hydrogen Ion. *Astrophys. J.* **1945**, *102*, 223–231. [CrossRef]
25. Chandrasekhar, S.; Elbert, D.D. On the Continuous Absorption Coefficient of the Negative Hydrogen Ion, V. *Astrophys. J.* **1958**, *128*, 633–635. [CrossRef]
26. Zanstra, H. A simple approximate formula for the recombination coefficient of hydrogen. *Observatory* **1954**, *74*, 66.
27. Geltman, S. The Bound-Free Absorption Coefficient of the Hydrogen Negative Ion. *Astrophys. J.* **1962**, *136*, 935–945. [CrossRef]
28. Doughty, N.A.; Fraser, P.A.; McEachran, R.P. The bound-free absorption coefficient of the negative hydrogen ion. *MNRAS* **1966**, *132*, 255–266. [CrossRef]
29. John, T.L.; Seaton, M.J. The Limiting Behaviour of the Absorption Cross Sections of the Negative Hydrogen Ion. *MNRAS* **1966**, *133*, 447–448. [CrossRef]
30. Bell, K.L.; Kingston, A.E. The bound-free absorption coefficient of the negative hydrogen ion. *Proc. Phys. Soc.* **1967**, *90*, 895–899. [CrossRef]
31. Matese, J.J.; Oberoi, R.S. Choosing Pseudostates in the Close-Coupling Formalism for the Electron—Atomic-Hydrogen System. *Phys. Rev. A* **1971**, *4*, 569–579. [CrossRef]
32. Hyman, H.A.; Jacobs, V.L.; Burke, P.G. Photoionization of H^- and He above the n=2 threshold. *J. Phys. B At. Mol. Phys.* **1972**, *5*, 2282–2291. [CrossRef]
33. Ajmera, M.P.; Chung, K.T. Photodetachment of negative hydrogen ions. *Phys. Rev. A* **1975**, *12*, 475–479. [CrossRef]

34. Broad, J.T.; Reinhardt, W.P. One- and two-electron photoejection from H^-: A multichannel J-matrix calculation *Phys. Rev. A* **1976**, *14*, 2159–2173. [CrossRef]
35. Reed, K.J.; Zimmerman, A.H.; Andersen, H.C.; Brauman, J.I. Cross sections for photodetachment of electrons from negative ions near threshold. *J. Chem. Phys.* **1976**, *64*, 1368–1375. [CrossRef]
36. Nascimento, M.A.C.; Goddard, W.A., III. The photodetachment cross section of the negative hydrogen ion. *Phys. Rev. A* **1977**, *16*, 1559–1567. [CrossRef]
37. Stewart, A.L. A perturbation-variation study of photodetachment from H^-. *J. Phys. B At. Mol. Phys.* **1978**, *11*, 3851–3861. [CrossRef]
38. Daskhan, M.; Roy, K.; Ghosh, A.S. Photodetachment cross section of negative hydrogen. *Indian J. Phys.* **1979**, *53B*, 183–189. [CrossRef]
39. Daskhan, M.; Ghosh, A.S. Photodetachment cross section of the negative hydrogen ion. *Phys. Rev. A* **1983**, *28*, 2767–2769. [CrossRef]
40. Wishart, A.W. The bound-free photodetachment cross section of H^-. *J. Phys. B At. Mol. Phys.* **1979**, *12*, 3511–3519. [CrossRef]
41. Esaulov, V.A. Electron detachment from atomic negative ions. *Ann. Phys. Fr.* **1986**, *11*, 493–592. [CrossRef]
42. Saha, H.P. Multiconfiguration Hartree-Fock calculation for the bound-free photodetachment cross section of H^-. *Phys. Rev. A* **1988**, *38*, 4546–4551. [CrossRef] [PubMed]
43. Venuti, M.; Decleva, P. Convergent multichannel continuum states by a general configuration interaction expansion in a B-spline basis: Application to H^- photodetachment. *J. Phys. B At. Mol. Opt. Phys.* **1997**, *30*, 4839-4859. [CrossRef]
44. Kuan, W.H.; Jiang, T.F.; Chung, K.T. Photodetachment of H^-. *Phys. Rev. A* **1999**, *60*, 364–369. [CrossRef]
45. Ivanov, V.K. Theoretical studies of photodetachment. *Radiat. Phys. Chem.* **2004**, *70*, 345–370. [CrossRef]
46. Frolov, A.M. Theoretical studies of photodetachment. *J. Phys. B At. Mol. Opt. Phys.* **2004**, *37*, 853. [CrossRef]
47. Morita, M.; Yabushita, S. Calculations of photoionization cross-sections with variationally optimized complex Gaussian-type basis functions. *Chem. Phys.* **2008**, *349*, 126–132. [CrossRef]
48. Kar, S.; Ho, Y.K. Photodetachment of the hydrogen negative ion in weakly coupled plasmas. *Phys. Plasmas* **2008**, *15*, 013301. [CrossRef]
49. Oanaa, C.M.; Krylov, A.I. Cross sections and photoelectron angular distributions in photodetachment from negative ions using equation-of-motion coupled-cluster Dyson orbitals. *J. Chem. Phys.* **2009**, *131*, 124114. [CrossRef]
50. Ghoshal, A.; Ho, Y.K. Photodetachment of H^-. *Phys. Rev. E* **2010**, *81*, 016403. [CrossRef]
51. Frolov, A.M. On the absorption of radiation by the negatively charged hydrogen ion. I. General theory and construction of the wave functions. *arXiv* **2013**, arXiv:1110.3432v7.
52. Ward, S.J.; McDowell, M.R.C.; Humberston, J.W. The Photodetachment of the Negative Ion of Positronium (Ps^-). *Europhys. Lett.* **1986**, *1*, 167–171. [CrossRef]
53. Hylleraas, E.A. Neue Berechnung der Energie des Heliums im Grundstande, sowie des tiefsten Terms von Ortho-Heliium. *Z. Physik* **1929**, *54*, 347–366 . [CrossRef]
54. Thakkar, A.J. Ph.D. Thesis (Queen's Univ. 1976). Available online: https://dl.dropboxusercontent.com/u/40687383/ThakkarPhD.pdf (accessed on 14 December 2014).
55. Landau, L.D.; Lifshitz, E.M. *Quantum Mechanics (Non-relativistic Theory), Course of Theoretical Physics Volume 3*; Pergamon Press: Oxford, UK, 1977; p. 606.
56. Drake, G.W.F. Electron-Hydrogen Photoattachment as a Source of Ultraviolet Absorption. *Ap. J.* **1974**, *189*, 161–163. [CrossRef]
57. Jacobs, V.L.; Bhatia, A.K.; Temkin, A. Photodetachment and Radiative Attachment Involving the 2PE3PE State of H^-. *Astrophys. J.* **1980**, *242*, 1278–1281. [CrossRef]
58. Ley-Koo, E.; Bunge, C.F. General evaluation of atomic electron-repulsion integrals in orbital methods without using a series representation for r_{12}^{-1}. *Phys. Rev. A* **1989**, *40*, 1215–1219. [CrossRef] [PubMed]
59. Edmonds, A.R. *Angular Momentum in Quantum Mechanics*; Princeton University Press: Princeton, NJ, USA, 1957; p. 124.
60. Joachain, C.J. *Quantum Collision Theory*; Elsevier Science Ltd.: Amsterdam, The Netherlands, 1983.
61. Prudnikov, A.P.; Brychkov, Y.A.; Marichev, O.I. *Integrals and Series*; Gorden and Breach: New York, NY, USA, 1986; Volume 3, p. 594, No. 7.13.1.1.

62. Luke, Y.L. *The Special Functions and their Approximations*; Academic Press: Cambridge, MA, USA, 2012; Volume 1, p. 212, No. 6.2.7.1.
63. Available online: http://functions.wolfram.com/03.01.26.0002.01 (accessed on 4 March 2020).
64. Bateman, H. *Higher Transcendental Functions*; McGraw-Hill: New York, NY, USA, 1953; Volume 1, p. 187, No. 4.3.14.
65. Available online: http://functions.wolfram.com/07.17.16.0002.01 (accessed on 4 March 2020).
66. Gradshteyn, I.S.; Ryzhik, I.M. *Table of Integrals, Series, and Products*, 5th ed.; Academic: New York, NY, USA, 1994.
67. We used the result Mathematica 7 gave, though it is likely one can get the same result from Prudnikov, A.P.; Brychkov, Y.A.; Marichev, O.I. *Integrals and Series*; Gorden and Breach: New York, NY, USA, 1986; Volume 3, p. 220, No. 2.21.2.11.
68. Available online: http://functions.wolfram.com/06.05.16.0001.01 (accessed on 4 March 2020).
69. Holzscheiter, M.H.; Charlton, M.; Nieto, M.M. The route to ultra-low energy antihydrogen. *Phys. Rep.* **2004**, *402*, 1–101. [CrossRef]
70. Bartmann, W.; Belochitskii, P.; Breuker, H.; Butin, F.; Marini de Carli, C.; Eriksson, T.; Maury, S.; Oelert, W.; Pasinelli, S.; Tranquille, G.A. Past, present and future low energy antiproton facilities at CERN. *Int. J. Mod. Phys. Conf. Ser.* **2014**, *30*, 1460261. [CrossRef]

 atoms

Review

A Precis of Threshold Laws for Positron vs. Electron Impact Ionization of Atoms

A. Temkin

Heliophysics Science Division, NASA/Goddard Space Flight Center, Greenbelt, MD 20771, USA; aaron.temkin-1@nasa.gov

Received: 21 February 2020; Accepted: 1 April 2020; Published: 10 April 2020

Abstract: The Coulomb-dipole theory of positron vs. electron impact ionization of hydrogen (as a proxy for neutral atoms) is reviewed, emphasizing how the analytic form of the threshold law (but not the magnitude) can be the same, whereas the physics of each is entirely different.

Keywords: Coulomb-dipole theory; positron vs. electron impact ionization

---

Threshold laws have been examined over a long period of time. It is not the purpose of this note to review the literature, but rather to concentrate on our own [1,2] approach and, in particular, to show how it applies to both positron as well as electron impact ionization, and to compare it with results from other Wannier [3] (electron) and modified Wannier (positron) approaches [4] to threshold laws for positron vs. electron impact ionization.

We start with the basic formula for the yield of two particles emerging from an initial state (Φ_i):

$$Q(E) = \int\int \|ME\|^2 \delta(E - \vec{k}_1{}^2 - \vec{k}_2{}^2) d^3k_1 d^3k_2 \quad (1)$$

where the transition matrix element is

$$ME = \int\int \Psi_f(\vec{k}_1, \vec{k}_2, \vec{r}_1, \vec{r}_2)[V_t(\vec{r}_1, \vec{r}_2) - V_i(\vec{r}_1, \vec{r}_2)]\Phi_i(\vec{r}_1, \vec{r}_2) d^3r_1 d^3r_2 \quad (2)$$

The ME is expressed in final state form, whereby Ψ_f is, in principle, the exact solution of the Schrodinger equation, (potential V_t); Ψ_f is normalized as a plane (s-partial) wave with unit amplitude at infinity, and Φ_i is the initial state before the interaction has taken place. Let us deal with electron/positron impact ionization of hydrogen. Then, the (s-wave) initial state is

$$\Phi_i = \frac{\sin(k_i r_1)}{k_i r_1} \phi_{10}(r_2) \quad (3)$$

Which implies (in Rydberg units)

$$V_i = -\frac{2}{r_2} \quad (4)$$

So that

$$Vt - Vi = \mp\frac{2}{r_1} \pm \frac{2}{r_{12}} \quad (5)$$

where the upper sign refers to electron and the lower sign positron scattering. Thus, the ME becomes explicitly

$$ME = \mp\int\int \Psi_f{}^{\mp}(\vec{r}_1, \vec{r}_2, \vec{k}_1, \vec{k}_2)(\frac{2}{r_1} - \frac{2}{r_{12}})\frac{\sin(k_i r_1)}{k_i r_1}\phi_{10}(r_2) d^3r_1 d^3r_2 \quad (6)$$

This shows explicitly how the ME, from which the threshold law is derived, relates to the exact solution, Ψ_f, of the Schrodinger equation. The Wannier law [3] does not deal directly with the wave function, but is (in my opinion) a brilliant analysis of the final state treated as a classical system. For electron-atom (ion) ionization, it gives the well-known threshold law

$$Q_-^{(W)}(E) \propto E^{-\frac{1}{4}+\frac{\mu}{2}}|_{\mu=\frac{1}{2}[\frac{100-9}{4z-1}]^{\frac{1}{2}}} \propto E^{1.1268} \tag{7a}$$

where z = charge of the residual ion (z = 1 for neutral atom, specifically H, ionization). For positron impact, the classical theory is quite different, and Klar [4] has derived

$$Q_+(E) \propto E^{2.65...} \tag{7b}$$

However, the classical assumption is not, in my opinion, fully justified. We believe the threshold is controlled by the region of phase space in which the outgoing particles do not share the available energy equally (i.e., the symbol ⊳ means between a factor 2 and 10). In this region, the faster particle sees the slower particle plus the residual ion as a unit, which is a dipole, whose dipole moment is the distance of the slower particle from the residual ion. At the same time, the slower particle sees the residual ion directly, i.e., it is a pure Coulomb wave. Thus, the name Coulomb-dipole (CD) theory. Mathematically, this derives from the fact that the potential reduces as below in the CD region

$$V_t = \mp\frac{2}{r_1} - \frac{2}{r_2} \pm \frac{2}{r_{12}} \cong -\frac{2}{r_2} \pm \frac{2r_2}{r_1^2}\cos(\vartheta_{12}) \tag{8}$$

For electron/positron impact, $\vartheta_{12} = \pi/0$. This says in the electron case that the two outgoing electrons repel each other and come out on opposite sides of the residual ion (the proton in the case of hydrogen), but with one electron much farther out than the inner electron, such that it sees the inner electron and the residual ion as a dipole facing away from the faster, outgoing electron, whereas in the positron impact case the outgoing positron is the faster particle and it sees the dipole formed by the slower electron and the residual ion facing toward the faster particle. In both cases, the potential between the faster particle and the dipole is attractive; thus,

$$V_t - V_i \cong -\frac{2r_2}{r_1^2} \tag{9}$$

Thus, the dipole moment seen by the faster particle is = $2r_2$. At the same time, the slower particle (an electron in both cases) comes out seeing the full charge of the residual ion (as a pure Coulomb wave). The associated wave function is

$$\Psi_f \propto F^d(r_1, r_2, k_1)F^c(r_2, k_2) \tag{10}$$

This is so in both cases. Thus, the matrix element reduces to

$$ME_{CD}(E, k_1, k_2) \propto \int_0^\infty r_1{}^2 dr_1 \int_0^{\frac{1}{2}r_1} r_2{}^2 dr_2 F^d(r_1,\ r_2, k_1)F^c(r_2, k_2)[\mp\frac{2}{r_1} \pm \frac{2}{r_1 \pm r_2}]\frac{\sin(k_i r_1)}{k_1 r_1}\phi_{10}(r_2) \tag{11}$$

In (11), we have switched to the SSCL (Spherically Symmetric Contra/Co Linear) model [5], wherein the interaction $\frac{2}{r_{12}} \rightarrow \frac{2}{r_1 \mp r_2}$, and in the CD region ($r_1 \rhd r_2$; $k_1 \rhd k_2$) the interaction in (11) reduces to $-\frac{2r_2}{r_1^2}$ (in the first approximation) in both cases (i.e., an attractive dipole). Thus, finally the CD matrix element reduces to

$$ME_{CD}(E,k_1,k_2) \propto \int_0^{\infty} r_1^2 dr_1 \int_0^{\frac{1}{2}r_1} r_2^2 dr_2 F^d(r_1,r_2k_1)F^c(r_2,k_2)[-\frac{2r_2}{r_1^2}]\frac{\sin(k_i r_1)}{k_1 r_1}\phi_{10}(r_2) \quad (12)$$

From (12), it is clear that both the electron and positron threshold laws will have the same form. The derivation of the threshold law from (12) via Equation (1) is given in Ref. [1,2]; it is

$$Q_{\pm}(E) \propto \frac{E}{(\ln(E))^2}[1 + C\sin(\alpha\ln(E) + \beta)] \quad (13)$$

Note first that (13) is not a pure power law. However, it is important also to realize its limitations, which are contained in (12) where the proportionality constant and limits of integration imply:

(a) The constant of proportionality as between positron vs. electron impact ionization will be entirely different. (The respective Schrodinger equations are different. It is only in the respective CD regions that the wave functions have a similar form, cf. Equation (10); I would expect the electron to be much larger than the positron constant of proportionality.)

(b) The energy distribution cross section, from which (13) is derived does not include the non-CD region $k_1 \geq k_2 \geq \frac{1}{2}k_1$. This means a major part of the energy distribution (cross section) is not included. Specifically, when ⊳ is interpreted > 2, we find for $k_2^2 > (1/5)E$ that the energy distribution is not covered by the CD theory. The part that is covered has the following form [4]:

$$\sigma_E(\varepsilon) \propto \frac{[1 + \cos(\alpha\ln(\varepsilon) + \beta))}{(\ln(\varepsilon))^2}] \quad (14)$$

In (14), ε is the energy of the slower electron in the region $\varepsilon < E/5$. Nevertheless, the area under this restricted part of the energy distribution curve is, according to the CD theory [1], larger than the area under the middle part, which is the dominant contributor to the total yield curve in the threshold limit $(E \rightarrow 0)$ [6,7].

We conclude by noting that a recent numerical calculation of the positron-atom impact cross section at very low energies by Bray et al. [8] gives results that are consistent with Klar's prediction. On the other hand, an experiment on positron-argon ionization [9], with an energy resolution not fine enough to test the modulation of the CD theory, found linear dependence, which is consistent with the CD theory [1,2] in the first approximation.

Funding: This research received no external funding.

Conflicts of Interest: The authors declare no conflict of interest.

References

1. Temkin, A. Threshold Law for Electron-Atom Impact Ionization. *Phys. Rev. Lett.* **1982**, *49*, 365. [CrossRef]
2. Temkin, A.; Shertzer, J. Electron scattering from excited states of hydrogen: Implications for the ionization threshold law. *Phys. Rev. A* **2013**, *87*, 052718. [CrossRef]
3. Wannier, G.H. The threshold law for single ionization of atoms or ions by electron impact. *Phys. Rev.* **1953**, *90*, 817. [CrossRef]
4. Klar, H. Threshold ionization of atoms by positrons. *J. Phys. B* **1981**, *14*, 4165. [CrossRef]
5. Temkin, A.; Hahn, Y. Optical potential approach to the e- -atom impact ionization threshold problem. *Phys. Rev.* **1974**, *10*, 708. [CrossRef]
6. Temkin, A. The Coulomb-Dipole Theory of the $e^{+/-}$ -Atom Impact Ionization Threshold Law. In *Electronic and Atomic Collisions*; North-Holland: Amsterdam, The Netherlands, 1983; p. 755.

7. Klar, H. *Electronic and Atomic Collisions*; North-Holland: Amsterdam, The Netherlands, 1984; p. 767.
8. Bray, I.; Bray, A.W.; Fursa, D.V.; Kadyrov, A.S. Near-Threshold Cross Sections for Electron qnd Positron Impact Ionization of Atomic Hydrogen. *Phys. Rev. Lett.* **2018**, *121*, 20341. [CrossRef]
9. Babij, T.J.; Machacek, J.R.; Murtagh, D.J.; Buckman, S.J.; Sullivan, J.P. Near thresholds ionization of argon by positrons. *Phys. Rev. Lett.* **2018**, *120*, 113401. [CrossRef] [PubMed]

Article

Positron Impact Excitation of the nS States of Atomic Hydrogen

Anand K. Bhatia

Heliophysics Science Division, NASA/Goddard Space Flight Center, Greenbelt, MD 20771, USA; Anand.K.Bhatia@nasa.gov

Received: 3 January 2020; Accepted: 26 February 2020; Published: 3 March 2020

Abstract: The excitation cross sections of the nS states, $n = 2$ to 6, of atomic hydrogen at various incident positron energies (10.23 to 300 eV) were calculated using the variational polarized-orbital method. Nine partial waves were used to obtain converged cross sections. The present results should be useful for comparison with results obtained from other theories and approximations. The positron-impact cross section was found to be higher than the electron-impact cross sections. Experimental and other theoretical results are discussed. The threshold law of excitation is discussed and the cross sections in this region were seen to obey the threshold law proportional to $(\ln k_f)^{-2}$. Cross sections were calculated in the Born approximation also and compared to those obtained using the variational polarized orbital method.

Keywords: positron-impact excitation; variational polarized orbital method; Born approximation

---

1. Introduction

Dirac [1] in 1928 formulated the well-known relativistic wave equation and predicted an antiparticle of spin $\hbar/2$. This antiparticle, later on called the positron, was detected experimentally by Anderson [2] by observing, in a cloud chamber, decay of cosmic ray pions into positrons and neutrinos. Over the years, research has been carried out to study its interaction with matter. For example, positronium annihilation has been observed from the center of the galaxy and also in solar flares. Positron annihilation can be used to study metallic defects [3]. They are essential in the formation of antihydrogen and in understanding of positron binding to neutral systems. Resonances in electron-atoms and electron-ions are ubiquitous but not in the case of positrons. However, some resonances have been calculated in positron-atom systems. Positrons are also used to detect diseases in a body by observing the positronium decay, called positron emission tomography.

The static potential between an incoming positron and a fixed target and the polarization potential are of opposite signs. The resultant potential is attractive but not attractive enough to bind a positron to atomic hydrogen. Total cross sections for positrons colliding with hydrogen atoms have been measured by Zhou et al. [4] in the energy range 1 to 1000 eV. Their measured cross sections include elastic scattering, excitation to all the higher levels, and perhaps ionization, but not annihilation and positronium cross sections. Total cross sections have been calculated by Gien [5] using the modified Glauber approximation, and by Walters [6] using a multi-pseudostate approximation, supplemented by the second Born approximation. He has calculated elastic and excitation of $1S$ to $2S$ and $2P$ cross sections, $2P$ cross sections being much larger than the $2S$ cross sections. Total cross sections calculated by him are the sum of elastic and excitation cross sections to all levels above $n = 1$. In a previous publication [7], results for positron-impact excitation cross sections for the excitation of the $2S$ state of atomic hydrogen were given for the incident positron energies (10.30 to 300 eV). It was shown that at high-incident energies cross sections are very close to those obtained using the Born approximation. In that calculation [7], there was a misprint in the computer program for scattering, found recently, affecting results for higher partial waves. For example, for $k = 0.90$ the cross section changes from previous value of 0.247 to 0.246, at $k = 1.2$ it changes from 0.338 to 0.317, and at the highest $k = 4.696$ it changes from 0.0128 to 0.0151. After correcting, the same calculation was repeated and extended to the excitation of higher states up to $n = 6$. Previous results were not much affected. Nine partial waves were used to obtain converged cross sections. The results of the previous calculation were compared with the close-coupling results of Burke et al. [8] and of Morgan [9] for the 2S excitation. A close-coupling calculation was carried out by Sarkar and Ghosh [10] with two basis sets of hydrogen and positronium states. The agreement is good with all these calculations and the comparison is not repeated here. There are various approximations to calculate excitation cross sections. The aim of this calculation is to provide another method for comparison of results obtained from various theories and approximations. Rydberg units are used and cross sections are in units of πa_0^2.

2. Theory and Calculations

The present calculations were carried out using the variational polarized-orbital method [11], using the expression for the cross section:

$$\sigma = \frac{k_f}{k_i}|T_{fi}|^2, \tag{1}$$

where k_i and k_f are the initial and final momenta of the positron, respectively, and the transition matrix is:

$$T_{fi} = -(\frac{1}{4\pi}) < \Psi_f \left| \frac{2}{r_1} - \frac{2}{r_{12}} \right| \Psi_i > \tag{2}$$

The positions of the incident positron and target electron is given by r_1 and r_2, $r_{12} = \left|\vec{r}_1 - \vec{r}_2\right|$. The nucleus is assumed to be of infinite mass and the incident wave function is Ψ_i, which, in principle, is an exact solution of the Schrödinger equation. It is given by:

$$\Psi_i(\vec{r}_1, \vec{r}_2) = u(\vec{r}_1)\Phi^{pol}(\vec{r}_1, \vec{r}_2) \tag{3}$$

The scattering function $u(\vec{r}_1)$ in the plane wave normalization for a partial wave L is given by:

$$u(\vec{r}_1) = a(L)\frac{u(r_1)}{r_1}Y_{L0}(\Omega_1). \tag{4}$$

The plane wave normalization is:

$$a(L) = \sqrt{4\pi(2L+1)}. \tag{5}$$

Other quantities in Equation (3) are given in [7] and the parameter β in Equation (8) of [7] is generally equal to 0.5. The final state wave function for a partial wave L is given by:

$$\Psi_f(\vec{r}_1, \vec{r}_2) = e^{i\vec{k}_f \cdot \vec{r}_1}\Phi_{nS}(\vec{r}_2) = 4\pi(i)^L j_L(k_f r)\sum Y_{Lm}(\hat{k}_f)Y_{Lm}(\Omega_1)\Phi_{nS}(\vec{r}_2) \tag{6}$$

Φ_{nS} are the excited S state functions. Using Equations (3) and (6) in Equation (2), we find that the cross section is given by:

$$\sigma_L(\pi a_0^2) = 16\pi(2L+1)^2\frac{k_f}{k_i}(Z_L)^2, \tag{7}$$

where

$$Z_L = \int r_1 dr_1 u(r_1) j_L(k_f r) X(r_1), \tag{8}$$

and

$$X(r_1) = \int d\vec{r}_2 \frac{Rns(r_2)}{r_{12}} (R_{1s}(r_2) + \frac{\chi(r_1)}{r_1^2} e^{-r_2}(0.5r_2^2 + r_2) \frac{\cos(\theta_{12})}{\sqrt{\pi}}). \quad (9)$$

j_L is the spherical Bessel function, R_{ns} and R_{1s} are the radial functions for the nS and $1S$ states. In the expression (7), we used:

$$\sum_m |Y_{Lm}(\hat{k}_f)|^2 = \frac{2L+1}{4\pi}, \quad (10)$$

Total cross sections converge when L is equal to 8 or less and it is given by:

$$\sigma(\pi a_0^2) = \sum_{L=0}^{8} \sigma_L(\pi a_0^2). \quad (11)$$

3. Born Approximation

The Born approximation is obtained by replacing the scattering function $u(\vec{r}_1)$ in Equation (3) by an incident plane wave $e^{i\vec{k}_i \cdot \vec{r}_i}$:

$$T_{fi} = -\frac{1}{4\pi} < e^{-i\vec{k}_f \cdot \vec{r}_1} \varphi_{2S}(\vec{r}_2) \left| \frac{2}{r_1} - \frac{2}{r_{12}} \right| e^{i\vec{k}_{\cdot i} \vec{r}_1} \Phi^{pol} > \quad (12)$$

$$T_{fi} = T_1 + T_2 \quad (13)$$

It can be shown that:

$$T_1 = \frac{4\sqrt{8}}{(a^2 + p^2)^3}, \quad (14)$$

$$T_2 = \frac{\sqrt{2}}{3} \int_0^\infty dr \sin(pr) \chi_{ST}(r) [-\frac{112}{r^2 a^5} + e^{-ar}(\frac{58r}{a^2} + \frac{56}{a^3} + \frac{112}{a^4 r} + \frac{112}{a^5 r^2})] \quad (15)$$

Cross section is given by:

$$\sigma = \frac{k_f}{k_i} \int d\widehat{k}_f \left| T \right|^2 \quad (16)$$

In the above equations, $\vec{p} = \vec{k}_i - \vec{k}_f$, a = 3/2, and χ_{ST} is given in Equation (7) of [7].

4. Results

Results for various cross sections, calculated in the variational polarized approximation [11], are given in Table 1 from k_i = 0.867 to k_i = 4.696. The higher excited state cross sections are small compared to the n = 2 excitation cross sections. However, they increase as the incident energy increases. Walters [6] has calculated elastic and excitation to $2S$ and $2P$ cross sections at k = 2.0, 2.711, 3.834, and 4.696 corresponding to 54.4, 100, 200, and 300 eV. In Table 2, results for elastic and total excitation cross sections obtained in this calculation are given and are compared to those obtained by Walters [6]. He has calculated elastic and excitation to $2S$ and $2P$ states, but not for higher states. The sum of these three cross sections does not add up to his total cross sections, which agree with those obtained by Gien [5] and the experimental results obtained by Zhou et al. [4]. It seems there is a substantial contribution from higher-excited nP, nD, and nF states. These calculations will be carried out in the near future.

Table 1. Excitation cross sections of nS states, $n = 1$ to 6, of atomic hydrogen using the variational polarized orbital method with $l_{max} = 8$.

k_i	2S	3S	4S	5S	6S	Total
0.867	4.68 (−2)					4.68 (−2)
0.8675	5.75 (−2)					5.75 (−2)
0.868	6.64 (−2)					6.64 (−2)
0.869	8.12 (−2)					8.12 (−2)
0.87	9.35 (−2)					9.35 (−2)
0.88	1.69 (−1)					1.69 (−1)
0.89	2.14 (−1)					2.14 (−1)
0.90	2.46 (−1)					2.46 (−1)
0.92	2.90 (−1)					2.90 (−1)
0.95	3.28 (−1)	8.90 (−4)				3.28 (−1)
0.96	3.35 (−1)	1.39 (−3)				3.37 (−1)
1.00	3.53 (−1)	2.60 (−3)	5.94 (−4)	2.32 (−2)	3.86 (−5)	3.79 (−1)
1.10	3.50 (−1)	4.27 (−3)	9.09 (−4)	1.02 (−1)	3.30 (−4)	4.57 (−1)
1.20	3.17 (−1)	5.20 (−3)	1.01 (−3)	4.71 (−2)	6.00 (−4)	3.71 (−1)
1.40	2.50 (−1)	6.03 (−3)	1.15 (−3)	5.40 (−2)	1.97 (−3)	3.13 (−1)
1.50	2.20 (−1)	6.12 (−3)	1.19 (−3)	5.62 (−2)	2.62 (−3)	2.86 (−1)
1.55	2.05 (−1)	6.18 (−3)	1.21 (−3)	5.72 (−2)	2.77 (−3)	2.72 (−1)
1.60	1.95 (−1)	6.19 (−3)	1.23 (−3)	8.13 (−2)	2.98 (−3)	2.63 (−1)
1.80	1.53 (−1)	6.07 (−3)	1.24 (−3)	5.91 (−2)	4.43 (−3)	2.24 (−1)
2.00	1.21 (−1)	5.83 (−3)	1.18 (−3)	5.68 (−2)	5.09 (−3)	1.90 (−1)
2.50	7.13 (−2)	5.11 (−3)	8.97 (−4)	4.36 (−2)	4.39 (−3)	1.25 (−1)
2.711	5.83 (−2)	4.80 (−3)	7.76 (−4)	3.77 (−2)	3.66 (−3)	1.05 (−1)
3.834	3.24 (−2)	3.30 (−3)	4.32 (−4)	2.05 (−2)	1.26 (−3)	5.85 (−2)
4.696	1.51 (−2)	2.41 (−3)	3.52 (−4)	1.66 (−2)	4.33 (−4)	3.49 (−2)

Table 2. Elastic and total excitation cross sections (πa_0^2) and those obtained by Walters.

E (eV)	A [a]	B [b]	C [c]	D [d]	E [e]	F [f]
54.4	3.54 (−1)	2.62 (−1)	6.16 (−1)	1.37	3.02	2.85
100.0	2.15 (−1)	1.05 (−1)	3.20 (−1)	9.80 (−1)	2.24	2.00
200.0	9.58 (−1)	5.85 (−2)	1.54 (−1)	6.28 (−1)	1.33	1.24
300.0	8.18 (−2)	3.49 (−2)	1.17 (−1)	4.65 (−1)	9.69 (−1)	8.73 (−1)

[a] Elastic cross sections. [b] Total excitation cross sections ($n = 2$ to 6).[c] Elastic and excitation cross sections in the present calculation. [d] Elastic and excitation to 2S and 2P states obtained by Walters [6]. [e] Total cross sections obtained by Walters [6]. [f] Total cross sections (interpolated) obtained by Gien [5].

In Figure 1, total excitation cross sections for 2S to 6S are shown. We added cross sections at $k_i = 0.867$, 0.8675, 0.868, and 0.869, close to the threshold. In the present calculation, very close to the threshold k_f tends to zero and the Bessel function is close to 1, making cross sections large in the threshold region.

The minimum is at $k = 0.87$ and the cross sections increase smoothly up to $k = 1.0$, then they start decreasing. The cross sections in the threshold region are shown in Figure 2. Wigner [12] has emphasized the importance of the long-range forces near the threshold region. The long-range force has been included in the present calculation, as indicated in [7,11]. In the threshold region the cross sections are proportional to $(\ln(k_f))^{-2}$ [13]. The cross sections calculated here in the threshold region can be fitted to $-0.03367 + \frac{3.477672}{(\ln(k_f))^2} - \frac{0.12238}{(\ln(k_f))^4}$, as shown by the solid line in Figure 2. Threshold behavior can be a useful diagnostic of the long-range potential [13].

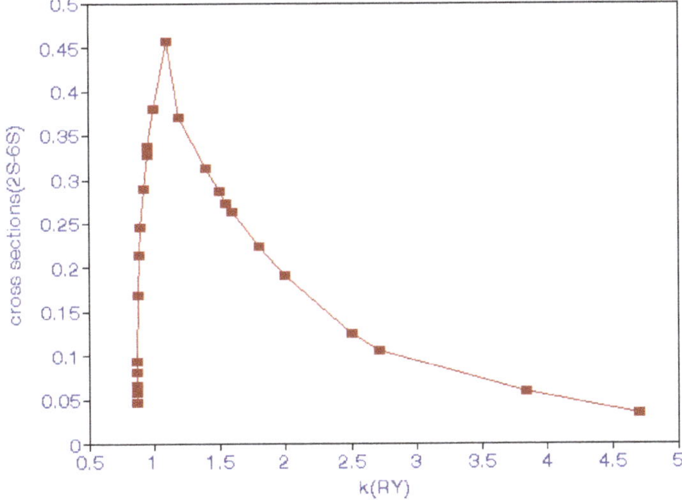

Figure 1. (Color online) Total excitation cross sections as a function of the incident k to nS states, $n = 2$ to 6.

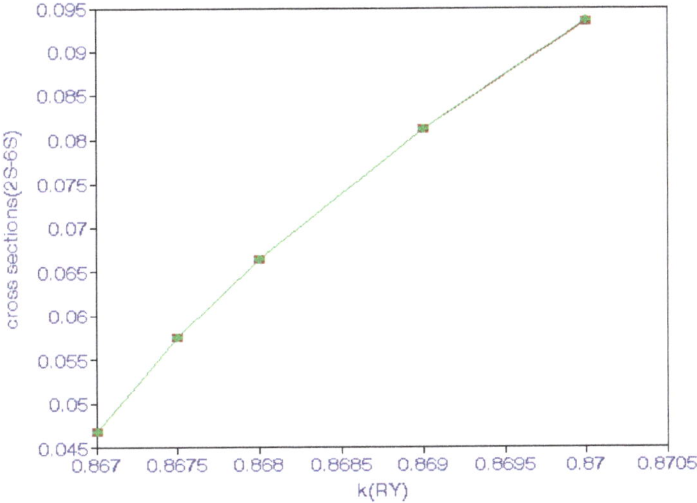

Figure 2. (Color online) $2S$ excitation cross sections as a function of the incident k in the threshold region, the solid line represents the fit to the squares (calculated cross sections).

No scaling law between the cross sections proportional to the power of n, the quantum number of the excited state, could be discerned. The cross sections including polarization of the target obtained in the Born approximation are given Table 3. In general the Born approximation is valid at high-incident energies. However, the agreement is quite good even at low-incident energies.

Table 3. 2S excitation cross sections in the Born approximation.

k_i	Cross Section	k_i	Cross Section	k_i	Cross Section
0.867	4.23 (−2)	0.92	3.11 (−1)	1.55	3.30 (−1)
0.8675	5.21 (−2)	0.95	3.77 (−1)	1.60	3.13 (−1)
0.868	6.03 (−2)	0.96	3.93 (−1)	1.80	2.53 (−1)
0.869	7.40 (−2)	1.00	4.42 (−1)	2.00	2.08 (−1)
0.87	8.56 (−2)	1.10	4.82 (−1)	2.711	1.15 (−1)
0.88	1.61 (−1)	1.20	4.66 (−1)	3.834	5.75 (−2)
0.89	2.10 (−1)	1.40	3.88 (−1)	4.696	3.83 (−2)
0.90	2.50 (−1)	1.50	3.49 (−1)		

A comparison of positron-impact excitation cross sections and electron-impact cross sections to the excited 2S is shown in Figure 3, which was given in the previous publication [7] and is repeated again because results at higher energies are not the same as before, as explained above. The positron-impact cross sections are higher than the electron-impact cross sections.

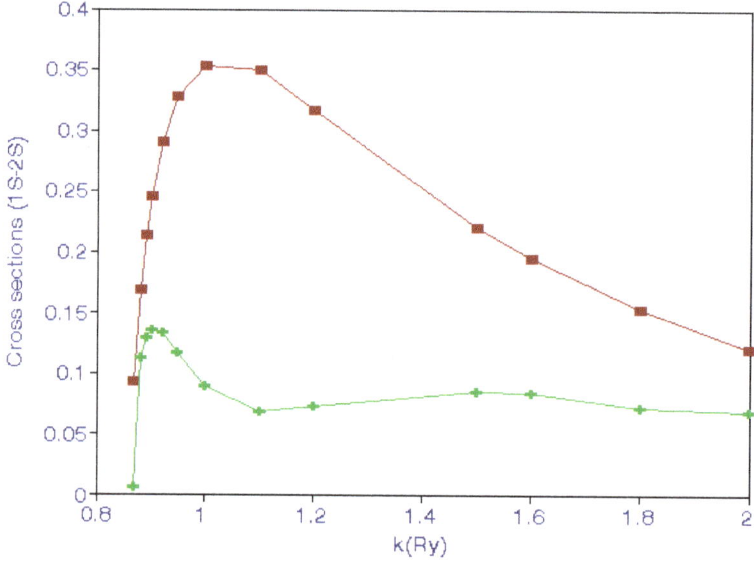

Figure 3. (Color online) Comparison of the 2S excitation cross sections as a function of the incident k of positron (upper curve) and electron impacts (lower curve).

5. Conclusions

We applied the variational polarized method to calculate excitation cross sections by positron impact for 2S state to 6S states. Results of this calculation were compared to the results obtained using other theories and approximations. The positron-impact cross sections were found to be higher than the electron-impact cross sections. Cross sections in the threshold region were proportional $(\ln(k_f)^{-2}$ and they were fitted to a form depending on $(\ln(k_f))^{-2}$ and $(\ln(k_f))^{-4}$. The present results were also compared with those obtained using the Born approximation.

Funding: The research received no external funding.

Acknowledgments: Thanks are extended to R. J. Drachman for helpful discussions and for critical reading of the manuscript.

Conflicts of Interest: The authors declare no conflicts of interest.

References

1. Dirac, P.A. *In the Development of Quantum Theory*; Gordon and Breach Science Publications: London, UK, 1971; p. 59.
2. Anderson, C.D. The positive electron. *Phys. Rev.* **1933**, *43*, 1933. [CrossRef]
3. Mackenzie, K.; Khooo, T.L.; McDonald, A.B.; Mckee, B.T.A. Temperature dependence of positron mean lives in metals. *Phys. Rev. Lett.* **1967**, *19*, 946. [CrossRef]
4. Zhou, S.; Kauppila, W.E.; Kwan, C.K.; Stein, T.S. Measurements of total cross sections for positrons and electrons colliding with atomic hydrogen. *Phys. Rev. Lett.* **1994**, *72*, 1443. [CrossRef] [PubMed]
5. Gien, T.T. Total cross sections for positron-hydrogen scattering. *J. Phys. B* **1995**, *28*, L321. [CrossRef]
6. Walters, H.R.J. Positron scattering by atomic hydrogen at intermediate energies: 1s to is, is to 2s and 1s to 2p transitions. *J. Phys. B* **1988**, *21*, 1893. [CrossRef]
7. Bhatia, A.K. Positron impact excitation of the 2*S* state of atomic hydrogen. *Atoms* **2019**, *7*, 69. [CrossRef]
8. Burke, P.G.; Schey, H.M.; Smith, K. Collision of slow electrons and positrons with atomic hydrogen. *Phys. Rev.* **1963**, *129*, 1258. [CrossRef]
9. Morgan, L.A. Positron impact excitation of the n = 2 level of hydrogen. *J. Phys. B At. Mol. Phys.* **1982**, *15*, L25–L29. [CrossRef]
10. Sarkar, N.K.; Ghosh, A.S. Positron hydrogen scattering at low and medium energies using CCA. *J. Phys. B* **1994**, *27*, 759–779. [CrossRef]
11. Bhatia, A.K. Hybrid theory of electron-hydrogen scattering. *Phys. Rev. A* **2007**, *75*, 032713. [CrossRef]
12. Wigner, E.P. On the behavior of cross sections near thresholds. *Phys. Rev.* **1948**, *73*, 1002. [CrossRef]
13. Sadeghpour, H.R.; Bohn, J.L.; Cavagnero, M.J.; Esry, B.D.; Fabrikant, I.I.; Macek, J.H.; Rau, A.R.P. Collisions near threshold in atomic and molecular physics. *J. Phys. B* **2000**, *33*, R93. [CrossRef]

Article

Calculations of Resonance Parameters for the Doubly Excited $^1P^o$ States in Ps^- Using Exponentially Correlated Wave Functions

Sabyasachi Kar [1,*] and Yew Kam Ho [2]

1 Department of Physics, Harbin Institute of Technology, Harbin 150001, China
2 Institute of Atomic and Molecular Sciences, Academia Sinica, Taipei 106, Taiwan; ykho@pub.iams.sinica.edu.tw
* Correspondence: skar@hit.edu.cn

Received: 28 November 2019; Accepted: 27 December 2019; Published: 31 December 2019

Abstract: Recent observations on resonance states of the positronium negative ion (Ps^-) in the laboratory created huge interest in terms of the calculation of the resonance parameters of the simple three-lepton system. We calculate the resonance parameters for the doubly excited $^1P^o$ states in Ps^- using correlated exponential wave functions based on the complex-coordinate rotation method. The resonance energies and widths for the $^1P^o$ Feshbach resonance states in Ps^- below the $N = 2, 3, 4, 5$ Ps thresholds are reported. The $^1P^o$ shape resonance above the $N = 2, 4$ Ps thresholds are also reported. Our predications are in agreement with the available results. Few Feshbach resonance parameters below the $N = 4$ and 5 Ps thresholds have been reported in the literature. Our predictions will provide useful information for future resonance experiments in Ps^-.

Keywords: positronium negative ion; Feshbach and shape resonance states; correlated exponential wave functions; complex-coordinate rotation method

1. Introduction

The study of doubly excited resonance states (DERS) in Ps^- has found significant relevance since the experimental observations of Michishio et al. [1] for a $^1P^o$ shape resonance in Ps^- near the $N = 2$ positronium (Ps) threshold. The DERS that appear from the closed channel and open channel segments of the scattering wave functions are commonly known, respectively, as Feshbach resonances (or closed channel resonances) and shape resonances (or open channel resonances). The present work aims to report on both the $^1P^o$ shape and Feshbach resonances in Ps^-.

To introduce this system, it would be of interest to recall its historical development from its existence and stability. The existence of the Ps^- was predicted by Wheeler in 1946 [2], and its ground state energy—the only stable state—was first reported by Hylleraas in 1946 [3]. Mills first reported the observation of the Ps^- in the laboratory [4]. He also reported the decay rate of this elusive ion [5]. Since then, a great number of theoretical studies and several experimental observations have been devoted to exploring the basic properties of this simplest bound three-lepton (e^-, e^+, e^-) system, and such a system was investigated as an interesting triatomic (XYX) molecule [6]. The molecular spectra of the Ps^- exhibiting the rotational and vibrational spectra are presented with illustrations in previous articles [6,7]. Theoretical predictions and experimental determinations for the Ps^- have been highlighted in recent papers [8–16].

The DERS in Ps^- were first reported by Ho [17]. He calculated the S-wave resonance parameters (RP) of this ion. After this pioneering work [17] on the resonance states in Ps^-, a great number of theoretical calculations on resonance states in Ps^- below the $N = 2$ Ps threshold have been reported in the literature, including the $^1P^o$ Feshbach [7,18–22] and shape [7,12,22,23] resonance states. Until now, many of these studies used different sophisticated methods or techniques or approaches, such as the technique of direct solution of the three-body scattering problem [24], the stabilization method (SM) [11,21,25,26], the complex-coordinate rotation method (CRM) [11–13,25,26], the technique of adiabatic molecular approximation [6], the Kohn variational method [27], the adiabatic treatment in hyperspherical coordinates (ATHC) [18,28,29], and the hyperspherical close-coupling approach (HCCA) [22,30–32].

First, we will briefly summarize previous works studying the doubly-excited $^1P^o$ resonance states, as these DERS are of our present interest. The $^1P^o$ Feshbach resonances in Ps^- below the $N = 2$ Ps threshold have been studied by Botero [18] using the ATHC, and by Bhatia and Ho [19], using the CRM with Hylleraas-type wave functions (HW). Ho and Bhatia [20] studied the doubly excited $^1P^o$ Feshbach resonance states below the $N = 3, 4, 5, 6$ Ps thresholds using the HW based on the CRM. The $^1P^o$ shape resonances in the Ps^- above the $N = 2, 4$ and 6 threshold have also been reported by Ho and Bhatia [23], using the HW and utilizing the CRM. Igarashi et al. [22] reported the $^1P^o$ Feshbach resonances near the $N = 2, 3, 4$ Ps thresholds and shape resonance associated with the $N = 2$ Ps threshold in the framework of HCCA. We have reported the $^1P^o$ Feshbach resonance parameter [21] below the $N = 2$ Ps threshold based on the SM and the $^1P^o$ shape resonance parameters [12] above the $N = 3$ Ps threshold based on the CRM using the exponentially correlated wave functions (ECW). In the present work, we calculate the $^1P^o$ Feshbach RP in Ps^- below the $N = 2, 3, 4, 5$ thresholds and the shape RP in Ps^- above the $N = 2$ and 4 thresholds by using the ECW and the CRM. Throughout this paper, the RP are meant for resonance energies and total widths, and atomic units (a.u.) are used unless stated otherwise.

2. Theory

The Hamiltonian (in atomic units) for the proposed (e^-, e^+, e^-) system can be written as

$$H = T + V, \tag{1}$$

$$T = -\frac{1}{2}\sum_{i=1}^{3} \nabla_i^2, \tag{2}$$

$$V = \sum_{\substack{i,j=1 \\ i<j}}^{3} \frac{q_i q_j}{r_{ij}}, \tag{3}$$

where q_1, q_2, and q_3 indicate the charges of two electrons 1, 2 and the positron, respectively and r_{ij} is the relative distance between the particle i and j.

As stated in the previous section, the $^1P^o$ state ECW can be proposed in the following form by introducing an overall scaling parameter ω and a permutation operator $\hat{P}_{12}$ for two electrons:

$$\Psi(\omega) = \sum_{\substack{i=1 \\ l_1+l_2=L}}^{N_B} C_i \varphi_i(\omega), \tag{4}$$

$$\varphi_i(\omega) = (1 + \hat{P}_{12}) \sum_{\substack{i=1 \\ l_1+l_2=L}}^{N_B} C_i \exp[(-\alpha_i r_{13} - \beta_i r_{23} - \gamma_i r_{21})\omega] \mathbf{Y}_{LM}^{l_1,l_2}(\mathbf{r}_{13}, \mathbf{r}_{23}) \tag{5a}$$

A well-known form of the bipolar harmonics $\mathbf{Y}_{LM}^{l_1,l_2}(\mathbf{r}_1,\mathbf{r}_2)$:

$$\mathbf{Y}_{LM}^{l_1,l_2}(\mathbf{r}_{13},\mathbf{r}_{23}) = r_{13}^{l_1} r_{23}^{l_2} \sum_{m_1,m_2} \begin{pmatrix} l_1 & l_2 & L \\ m_1 & m_2 & -M \end{pmatrix} Y_{l_1 m_1}(\hat{r}_{13}) Y_{l_1 m_2}(\hat{r}_{23}), \tag{5b}$$

where N_B is the number of basis terms. The nonlinear variational parameters $\alpha_i, \beta_i, \gamma_i$ in the ECW (4) are generated by the proper choice of a quasi-random process of the following form

$$Z_i = \left[\frac{1}{2}k(k+1)\sqrt{p_Z}\right](R_{2,Z} - R_{1,Z}) + R_{1,Z}, \tag{6}$$

$[z]$ assumes the fractional part of z, the real intervals $[R_{1,Z},\ R_{2,Z}]$ $(Z = \alpha, \beta, \gamma)$ require optimization to obtain the appropriate values of $R_{1,Z}$ and $R_{2,Z}$. p_Z stands for a prime number and it takes the numbers 2, 3, and 5 for $Z = \alpha$ $Z = \beta$ and $Z = \gamma$, respectively.

To set the present DERS calculations using the CRM [33], the radial coordinates are transformed following a dilation rule comprising of the so-called rotational angle θ.

$$r \to r\exp(i\theta), \tag{7}$$

and the form of the transformed Hamiltonian:

$$H(\theta) \to \left(-\frac{1}{2}\sum_{i=1}^{3} \nabla_i^2\right)\exp(-2i\theta) + \left(\sum_{\substack{i,j=1\\ i<j}}^{3} \frac{q_i q_j}{r_{ij}}\right)\exp(-i\theta), \tag{8}$$

In the case of nonorthogonal basis functions, the overlap and Hamiltonian matrices take the form

$$N_{nm}(\omega) = \langle \varphi_n(\omega) \big| \varphi_m(\omega) \rangle \tag{9}$$

and

$$H_{nm}(\theta,\omega) = \langle \varphi_n(\omega) \big| H(\theta) \big| \varphi_m(\omega) \rangle \tag{10}$$

The complex characteristic values can be obtained by solving the equation

$$\sum_n \sum_m C_{nm}[H_{nm}(\theta,\omega) - E(\theta,\omega)N_{nm}(\omega)] = 0 \tag{11}$$

Resonance poles can be located by observing the complex energy $E(\theta,\omega)$ for various values of θ and ω. The complex resonance energy is given by

$$E_{res} = E_r - \frac{i\Gamma}{2} \tag{12}$$

where E_r is the resonance energy, and Γ is the width. The RP are identified by locating stabilized roots with respect to the variation of the scaling parameter ω in the ECW for optimum choice of the nonlinear variational parameters $\alpha_i, \beta_i, \gamma_i$, and of the rotational angle θ.

3. Results and Discussions

To extract RP (resonance positions and widths), we calculate the complex-energy eigen values $E(\theta,\omega)$ for different values of θ and ω by diagonalizing the transformed Hamiltonian matrix. For the present problem, the parameters θ and ω are varied respectively from 0.00 to 0.60 with mesh size 0.02 and from 0.1 to 0.6 with mesh size 0.001. Exploiting the computational technique, we extract the $^1P^o$ resonance states associated with the N = 2, 3, 4, 5 Ps threshold. Figure 1 depicts the rotational paths near the pole for the $^1P^o$ (1) Feshbach resonance (FR) in the Ps^- lying below the Ps (N = 2) threshold in the energy plane (EP) for five different values of the scaling parameter ω using 900-term ECW. In this figure θ = 0.20 (002) 0.40 means the θ assumes the value from 0.2 to 0.4 with mesh size 0.02. Similarly, Figures 2–5 shows respectively the rotational paths near the poles for the $^1P^o$ (2) Feshbach resonance below the N = 2 Ps threshold, for the $^1P^o$ (4) FR below the N = 3 Ps threshold, for the $^1P^o$ (5) Feshbach resonance below the N = 5 Ps threshold, and for the $^1P^o$ (7) FR below the N = 5 Ps threshold. From Figure 1, we can determine the RP $(E_r, \Gamma/2)$ as $(-0.063155862 \pm 2 \times 10^{-9}, 0.459 \times 10^{-6} \pm 1 \times 10^{-9})$ a.u. In a similar way, from Figures 2–5, we can estimate the RP $(E_r, \Gamma/2)$ respectively as $(-0.06254245 \pm 4 \times 10^{-8}$, $0.11 \times 10^{-6} \pm 3 \times 10^{-8})$, $(-0.0281013 \pm 1 \times 10^{-6}, 0.169 \times 10^{-4} \pm 2 \times 10^{-6})$, $(-0.010580 \pm 3 \times 10^{-6}, 0.16 \times 10^{-4} \pm 3 \times 10^{-6})$, and $(-0.010384 \pm 3 \times 10^{-6}, 0.19 \times 10^{-4} \pm 3 \times 10^{-6})$. All the results obtained from the present calculations are summarized in Tables 1 and 2.

In Table 1, we present the $^1P^o$ Feshbach resonance energies and widths (in a.u.) in Ps^- below the N = 2, 3, 4, 5 Ps thresholds. The $^1P^o$ (7) FR below the N = 4 Ps threshold and the $^1P^o$ (5), $^1P^o$ (6), $^1P^o$ (7), $^1P^o$ (8) Feshbach resonances below the N = 5 Ps threshold are reported for the first time in the literature, to the best of our knowledge. In this table, we also present the $^1P^o$ shape resonances obtained from the resent calculations above the N = 2 and 4 Ps thresholds. The $^1P^o$ shape RP above the N = 3 threshold are taken from our recent work [12] and are presented in this table for completeness. In Table 1, we have also included the available results from the other calculations [18–23]. Table 1 shows that our predications of RP using the ECW are in agreement with the results of Bhatia et al. [19] and Ho et al. [20,23] using HW and the CRM. The $^1P^o$ intrashell resonance states are also in agreement with those obtained by Ivanov and Ho [7] using CI-type basis functions and the CRM. Table 1 also shows that our resonance energies and widths are fairly comparable with the reported results of Igarashi et al. [22], except for the resonance widths of the $^1P^o$ (4), $^1P^o$ (5), $^1P^o$ (6) resonance states below the N = 4 Ps threshold. The numbers in the table inside parentheses denote the uncertainty in the last digit. But our listed results in this table from the works of Ho and collaborators [7,19,20,23] are converted from Rydberg units to atomic units, and so the uncertainty in the last digit exhibits a value with fractional part. The discrepancy in resonance widths is probably due to the technique used or due to interference of higher lying states in HCCA calculations. The discrepancy with other calculations in terms of precision is probably due to the use of different computational tools in different calculations. Though only a few results are presented as new in this paper, all of the results shown in Table 1 were obtained using different wave functions.

Our estimated FR energy and width below the Ps (N = 2) threshold are also in good accord with our previous work using the 600-term CEW and the stabilization method [21]. In Table 2, we present our calculated resonance energies for the doubly excited $^1P^o$ states using the ECW and the CRM in electron volts (eV). To express our present results from a.u. to eV, we measure the resonance energy from the ground state of the Ps^- (−0.2620050702325 a.u. [34]). The corresponding resonance widths obtained from this calculation are presented in meV. To convert a.u. to eV, we use the relation 1 a.u. = 27.21138501195 [35]. Table 2 shows that the $^1P^o$ shape resonance above the N = 2 Ps threshold obtained from the present calculation is in good agreement with the recent experimental observation [1]. We have examined convergence of our calculations with the increasing number of terms in ECW. We have also studied the stability of our works with different choices of nonlinear variational parameters. Our estimated resonance parameters are convergent and stable up to the quoted digits in Table 1.

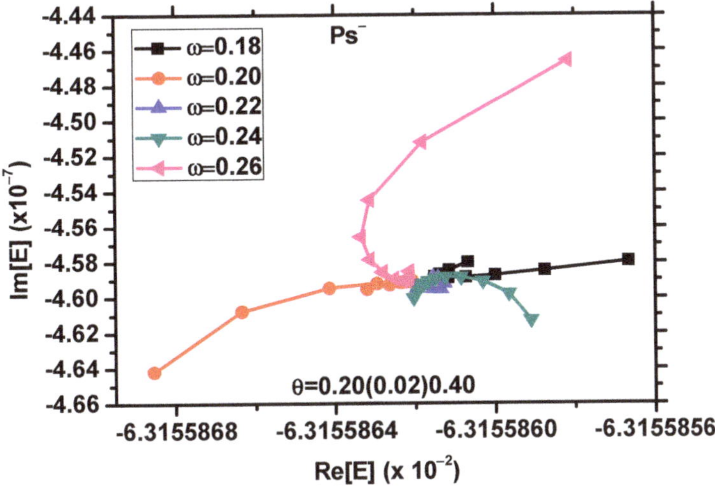

Figure 1. The rotational paths near the pole for the ${}^1P^{o}$ (1) FR of Ps^- lying below the Ps ($N = 2$) threshold in the EP for five different values of the scaling parameter ω using 900-term wave ECW.

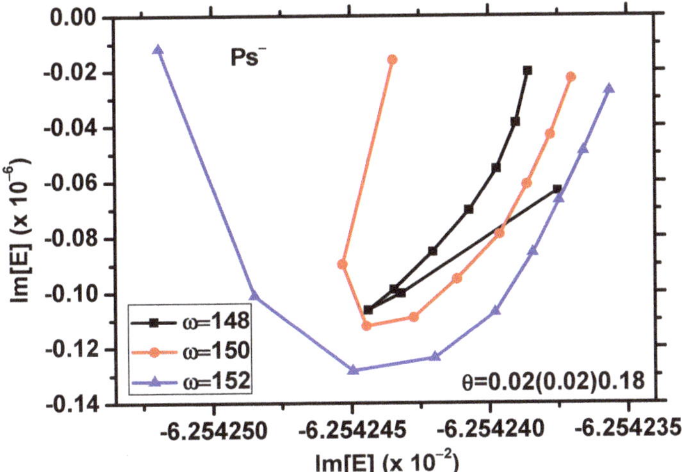

Figure 2. The rotational paths near the pole for the ${}^1P^{o}$ (2) FR of Ps^- lying below the Ps ($N = 2$) threshold in the EP for three different values of the scaling parameter ω using 900-term ECW.

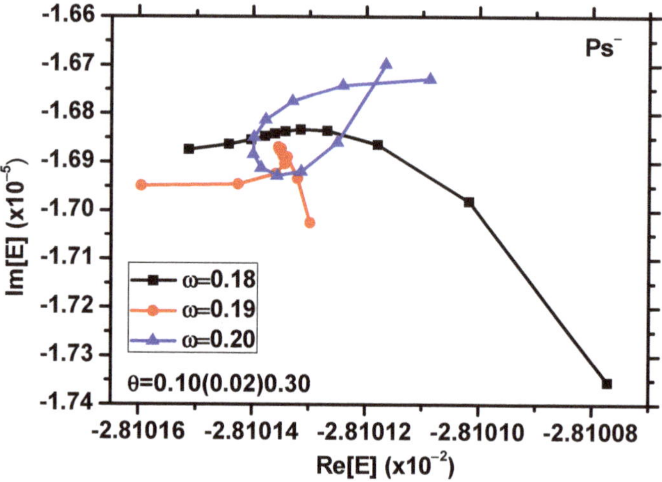

Figure 3. The rotational paths near the pole for the $^{1}P^{o}$ (4) FR of Ps^{-} lying below the Ps ($N = 3$) threshold in the EP for three different values of the scaling parameter ω using 900-term ECW.

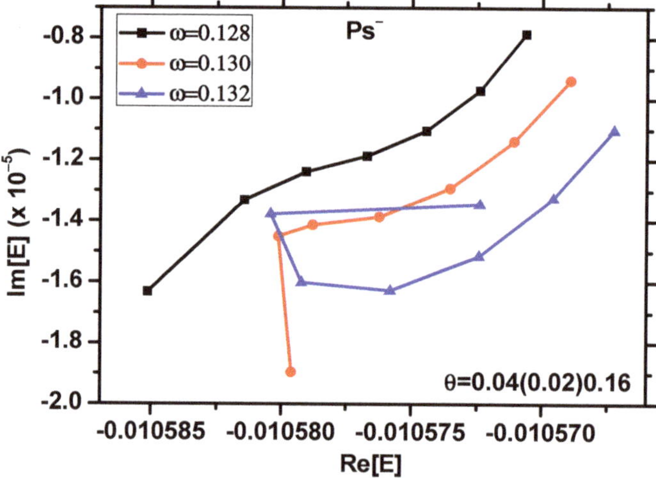

Figure 4. The rotational paths near the pole for the $^{1}P^{o}$ (5) FR of Ps^{-} lying below the Ps ($N = 5$) threshold in the EP for three different values of the scaling parameter ω using 900-term ECW.

Table 1. The $^1P^o$ Feshbach and shape RPs (in a.u.) in Ps^- associated with the $N = 2, 3, 4, 5$ Ps thresholds. The numbers in the parentheses denote the uncertainty in the last digit.

	Present Calculations		Other Calculations		
	E_r	$\frac{\Gamma}{2}$	E_r	$\frac{\Gamma}{2}$	Reference
			$N = 2$: $Eth = -0.0625$		
$^1P^o$ (1)	−0.063155862(2)	$0.459(1) \times 10^{-6}$	−0.0631553(1.5) −0.0631559 −0.063155 −0.0625087	$0.5(1.5) \times 10^{-6}$ 0.4435×10^{-6} 0.41×10^{-6}	a b c d
$^1P^o$ (2)	−0.06254244(4)	$0.11(3) \times 10^{-6}$	−0.062543	0.125×10^{-6}	c
$^1P^o$ (shape)	−0.06218(2)	0.00020(2)	−0.06217(1.5) −0.062158	0.000225(1.5) 0.00032	e c
			$N = 3$: $Eth = -0.0277777777778$		
$^1P^o$ (1)	−0.03162236(2)	0.0001103(2)	−0.03162235(0.5) −0.031621	0.0001103(0.5) 0.00011	a c
$^1P^o$ (2)	−0.02921495(2)	$0.75(2) \times 10^{-6}$	−0.02921495(0.5) −0.029212	$0.75(0.5) \times 10^{-6}$ 0.75×10^{-6}	a c
$^1P^o$ (3)	−0.0281276(1)	$0.33(3) \times 10^{-6}$	−0.028125	0.30×10^{-6}	c
$^1P^o$ (4)	−0.0281013(1)	$0.169(2) \times 10^{-4}$	−0.028099	0.165×10^{-4}	c
$^1P^o$ (5)	−0.027863(1)	$0.2(1) \times 10^{-6}$	−0.027864	0.435×10^{-7}	c
$^1P^o$ (6)	−0.027809(2)	$0.28(3) \times 10^{-5}$	−0.027811	0.175×10^{-5}	c
$^1P^o$ (shape)	−0.0255(2)	0.0021(2)			f
			$N = 4$: $Eth = -0.015625$		
$^1P^o$ (1)	−0.01889032(1)	0.0000154	−0.01889035(0.5) −0.018890385(1) −0.018863	0.0000154(0.5) 0.000015395 0.000016	g h c
$^1P^o$ (2)	−0.01704109(1)	$0.8(2) \times 10^{-6}$	−0.01704125(0.5) −0.017031	0.65×10^{-6} 0.55×10^{-6}	g c
$^1P^o$ (3)	−0.016536(2)	$0.10(2) \times 10^{-4}$	−0.0165385(0.5) −0.016480	$0.098(0.5) \times 10^{-4}$ 0.1×10^{-4}	g
$^1P^o$ (4)	−0.016163(1)	$0.2(1) \times 10^{-5}$	−0.016161(2.5) −0.016139	$0.235(2.5) \times 10^{-5}$ 0.21×10^{-6}	g c
$^1P^o$ (5)	−0.015882(2)	$0.20(2) \times 10^{-4}$	−0.015880 (2.5) −0.015855	$0.085(2.5) \times 10^{-4}$ 0.32×10^{-5}	g c
$^1P^o$ (6)	−0.015802(1)	$0.10(1) \times 10^{-5}$	−0.0158025(1) −0.015819	$0.125(1) \times 10^{-5}$ 0.65×10^{-7}	g c
$^1P^o$ (7)	−0.01566(2)	$0.035(3) \times 10^{-5}$			
$^1P^o$ (shape)	−0.01548(1)	0.000022(2)	−0.0154875(0.5) −0.0154775(1.5)	0.000015(0.5) 0.0000305	e h
			$N = 5$: $Eth = -0.01$		
$^1P^o$ (1)	−0.012463(2)	$0.16(2) \times 10^{-4}$	−0.0124625(0.5) −0.01246295	$0.1525(5) \times 10^{-4}$ 0.1525×10^{-4}	g h
$^1P^o$ (2)	−0.011216(1)	$0.2(1) \times 10^{-5}$	−0.0112155(2.5)	0.135×10^{-5}	g
$^1P^o$ (3)	−0.01104(1)	$0.22(1) \times 10^{-4}$	−0.01104375(0.5)	$0.1575(0.5) \times 10^{-4}$	g
$^1P^o$ (4)	−0.01083(2)	$0.70(2) \times 10^{-4}$	−0.010830(0.5) −0.01083009(1.5)	$0.68(0.5) \times 10^{-4}$ 0.68045×10^{-4}	g h
$^1P^o$ (5)	−0.010580(3)	$0.16(3) \times 10^{-4}$			
$^1P^o$ (6)	−0.01048(1)	$0.2(1) \times 10^{-4}$			
$^1P^o$ (7)	−0.010384(3)	$0.19(3) \times 10^{-4}$			
$^1P^o$ (8)	−0.01022(1)	$0.7(1) \times 10^{-4}$			

a: Bhatia and Ho [19], b: Kar and Ho [21], c: Igarashi et al. [22], d: Botero [18], e: Ho and Bhatia [23], f: Kar and Ho [12], g: Ho and Bhatia [20], h: Ivanov and Ho [7].

Table 2. The $^1P^\circ$ Feshbach and shape RP (in eV) in Ps^- associated with the $N = 2, 3, 4, 5$ Ps thresholds. The resonance positions are measured from the ground state of the Ps^- ion.

	Present Calculations	
	E_r(eV)	Γ (meV)
	$N = 2$	
$^1P^\circ$ (1)	5.4109623645	0.024980052
$^1P^\circ$ (2)	5.4276543	0.0060
$^1P^\circ$ (shape)	5.43752 5.437(1) [a]	10.88456 10(2) [a]
	$N = 3$	
$^1P^\circ$ (1)	6.26903263	6.002832
$^1P^\circ$ (2)	6.33454159	0.0408
$^1P^\circ$ (3)	6.36412989	0.01796
$^1P^\circ$ (4)	6.36484555	0.9198
$^1P^\circ$ (5)	6.37133002	0.0109
$^1P^\circ$ (6)	6.372799	0.1524
$^1P^\circ$ (shape)	6.4356 [b]	114.29 [b]
	$N = 4$	
$^1P^\circ$ (1)	6.61548907	0.83811
$^1P^\circ$ (2)	6.66580918	0.0436
$^1P^\circ$ (3)	6.6795534	0.544
$^1P^\circ$ (4)	6.6897032	0.108
$^1P^\circ$ (5)	6.6973496	1.088
$^1P^\circ$ (6)	6.6995265	0.0544
$^1P^\circ$ (7)	6.70339	1.9048
$^1P^\circ$ (shape)	6.70829	1.198
	$N = 5$	
$^1P^\circ$ (1)	6.790385	0.0870
$^1P^\circ$ (2)	6.824318	0.108
$^1P^\circ$ (3)	6.829107	1.20
$^1P^\circ$ (4)	6.834822	3.810
$^1P^\circ$ (5)	6.841624	0.870
$^1P^\circ$ (6)	6.844346	1.08
$^1P^\circ$ (7)	6.846958	1.034
$^1P^\circ$ (8)	6.85142	3.80

[a] Experiment (Ref. [1]), [b] Our recent work (Ref. [12]).

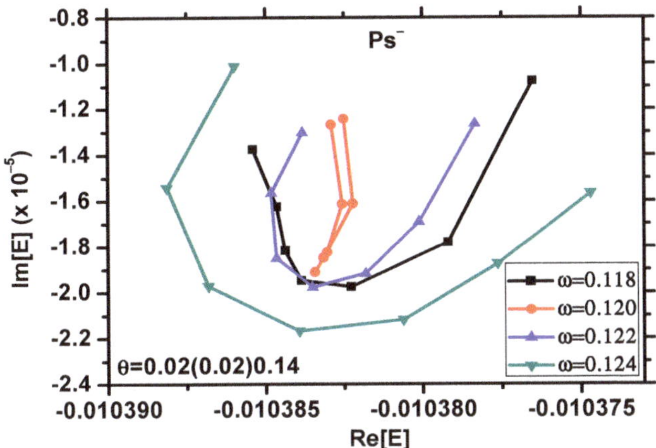

Figure 5. The rotational paths near the pole for the $^1P^o$ (7) FR of Ps^- lying below the Ps ($N = 5$) threshold in the EP for four different values of the scaling parameter ω using 900-term ECW.

4. Conclusions

In this work, we have calculated the resonance parameters for the doubly excited $^1P^o$ states in Ps^- using correlated exponential wave functions in the framework of the complex-coordinate rotation method. The resonance energies and widths for the $^1P^o$ resonance states in Ps^- below the $N = 2, 3, 4, 5$ Ps thresholds are reported. The $^1P^o$ shape resonance above the $N = 2, 4$ Ps thresholds are also reported. Few resonance states have been identified for the first time in the literature. The resonance energies and widths obtained from this work, using different wave functions as compared with those used in earlier investigations, are in agreement with the available data. With the recent experimental observation of the $^1P^o$ shape resonance states in the positronium ions, it is hoped that our investigations for the doubly excited $^1P^o$ resonance states will provide useful information for future resonance experiments in Ps^-.

Author Contributions: All the authors contributed equally in this work. All authors have read and agreed to the published version of the manuscript.

Funding: The research received no external funding.

Conflicts of Interest: The authors declare no conflicts of interest.

References

1. Michishio, K.; Kanai, T.; Kuma, S.; Azuma, T.; Wada, K.; Mochizuki, I.; Hyodo, T.; Yagishita, A.; Nagashima, Y. Observation of a shape resonance of the positronium negative ion. *Nat. Commun.* **2016**, *7*, 11060. [CrossRef] [PubMed]
2. Wheeler, J.A. Polyelectrons. *Ann. N. Y. Acad. Sci.* **1946**, *48*, 219–238. [CrossRef]
3. Hylleraas, E.A. Electron affinity of positronium. *Phys. Rev.* **1947**, *71*, 491. [CrossRef]
4. Mills, A.P., Jr. Observation of the positronium negative ion. *Phys. Rev. Lett.* **1981**, *46*, 717. [CrossRef]
5. Mills, A.P., Jr. Measurement of the decay rate of the positronium negative ion. *Phys. Rev. Lett.* **1983**, *50*, 671. [CrossRef]
6. Rost, J.M.; Wingten, D. Positronium negative ion: Molecule or atom? *Phys. Rev. Lett.* **1992**, *69*, 2499. [CrossRef]
7. Ivanov, I.A.; Ho, Y.K. Supermultiplet structures of the doubly excited positronium negative ion. *Phys. Rev. A* **2000**, *61*, 032501. [CrossRef]
8. Ho, Y.K. Atomic resonances involving positrons. *Nucl. Instrum. Method Phys. Res. B* **2008**, *266*, 516–521. [CrossRef]

9. Michishio, K.; Tachibana, T.; Terabe, H.; Igarashi, A.; Wada, K.; Kuga, T.; Yagishita, A.; Hyodo, T.; Nagashima, Y. Photodetachment of positronium negative ions. *Phys. Rev. Lett.* **2011**, *106*, 153401. [CrossRef]
10. Nagashima, Y. Experiments on positronium negative ions. *Phys. Rep.* **2014**, *545*, 95–123. [CrossRef]
11. Kar, S.; Ho, Y.K. Excitons and the positronium negative ion: comparison in spectroscopic properties. In emphExcitons; Pyshkin, S.L., Ed.; INTECH: London, UK, 2018; Chapter 5; pp. 69–90.
12. Kar, S.; Ho, Y.K. A new ^{1}P° shape resonance in Ps$^-$ above the Ps ($N = 3$) threshold. *Phys. Lett. A* **2018**, *382*, 1787–1790. [CrossRef]
13. Kar, S.; Ho, Y.K. Two-photon double-electron D-wave resonant excitation in the positronium negative ion. *Eur. Phys. J. D* **2018**, *72*, 193. [CrossRef]
14. Cassidy, D.B. Experimental progress in positronium laser physics. *Eur. Phys. J. D* **2018**, *72*, 53. [CrossRef]
15. Kar, S.; Wang, Y.S.; Wang, Y.; Ho, Y.K. Polarizability of negatively charged helium-like ions interacting with Coulomb and screened Coulomb potentials. *Int. J. Quantum Chem.* **2018**, *118*, e25515. [CrossRef]
16. Kar, S.; Wang, Y.S.; Wang, Y.; Ho, Y.K. Critical stability of the negatively charged positronium like ions with Yukawa potentials and varying Z. *Atoms* **2019**, *7*, 53. [CrossRef]
17. Ho, Y.K. Autoionization states of the positronium negative ion. *Phys. Rev. A* **1979**, *19*, 2347. [CrossRef]
18. Botero, J. Adiabatic study of the positronium negative ion. *Phys. Rev. A* **1987**, *35*, 36–50. [CrossRef]
19. Bhatia, A.K.; Ho, Y.K. Complex-coordinate calculation of ^{1,3}P resonances in Ps$^-$ using Hylleraas functions. *Phys. Rev. A* **1990**, *42*, 1119–1122. [CrossRef]
20. Ho, Y.K.; Bhatia, A.K. ^{1,3}P° resonance states in positronium ions. *Phys. Rev. A* **1991**, *44*, 2890–2894. [CrossRef]
21. Kar, S.; Ho, Y.K. Doubly excited ^{1,3}P° resonance states of Ps$^-$ in weakly coupled plasmas. *Phys. Rev. A* **2006**, *73*, 032502. [CrossRef]
22. Igarashi, A.; Shimamura, I.; Toshima, N. Photodetachment cross sections of the positronium negative ion. *New J. Phys.* **2000**, *2*, 17. [CrossRef]
23. Ho, Y.K.; Bhatia, A.K. P–wave shape resonances in positronium ions. *Phys. Rev. A* **1993**, *47*, 1497–1499. [CrossRef] [PubMed]
24. Hu, C.-Y.; Kvitsinsky, A.A. Resonances in e$^-$-Ps elastic scattering via a direct solution of the three-body scattering problem. *Phys. Rev. A* **1994**, *50*, 1924–1926. [CrossRef] [PubMed]
25. Kar, S.; Ho, Y.K. The ^{1,3}D° resonance states of positronium negative ion using exponential correlated wave functions. *Eur. Phys. J. D* **2010**, *57*, 13–19. [CrossRef]
26. Kar, S.; Ho, Y.K. Resonance states of Ps$^-$ using correlated wave functions. *Comput. Phys. Commun.* **2011**, *181*, 119–121. [CrossRef]
27. Ward, S.J.; Humberston, J.W.; McDowell, M.R.C. Elastic scattering of electrons (or positrons) from positronium and the photodetachment of the positronium negative ion. *J. Phys. B* **1987**, *20*, 127–149. [CrossRef]
28. Botero, J.; Greene, C.H. Resonant photodetachment of the positronium negative ion. *Phys. Rev. Lett.* **1986**, *56*, 1366–1369. [CrossRef]
29. Botero, J. Adiabatic hyperspherical study of three-particle systems. *Z. Phys. D* **1988**, *8*, 177–180. [CrossRef]
30. Zhou, Y.; Lin, C.D. Comparative Studies of Excitations and Resonances in H$^-$, Ps$^-$, and e$^+$ + H Systems. *Phys. Rev. Lett.* **1995**, *75*, 2296. [CrossRef]
31. Igarashi, A.; Shimamura, I. Time-delay matrix analysis of resonances: Application to the positronium negative ion. *J. Phys. B* **2004**, *37*, 4221. [CrossRef]
32. Aiba, K.; Igarashi, A.; Shimamura, I. Time-delay matrix analysis of several overlapping resonances: Applications to the helium atom and the positronium negative ion. *J. Phys. B* **2007**, *40*, F9–F17. [CrossRef]
33. Ho, Y.K. The method of complex coordinate rotation and its applications to atomic collision processes. *Phys. Rep.* **1983**, *99*, 1–68. [CrossRef]
34. Kar, S.; Ho, Y.K. Positron annihilation in plasma-embedded Ps$^-$. *Chem. Phys. Lett.* **2006**, *424*, 403–408. [CrossRef]
35. Mohr, P.J.; Newell, D.B.; Taylor, B.N. CODATA Recommended Values of the Fundamental Physical Constants: 2014*. *J. Phys. Chem. Ref. Data* **2016**, *45*, 043102. [CrossRef]

MDPI
St. Alban-Anlage 66
4052 Basel
Switzerland
Tel. +41 61 683 77 34
Fax +41 61 302 89 18
www.mdpi.com

Atoms Editorial Office
E-mail: atoms@mdpi.com
www.mdpi.com/journal/atoms

www.ingramcontent.com/pod-product-compliance
Lightning Source LLC
LaVergne TN
LVHW070634170726
843515LV00004B/247
* 9 7 8 3 0 3 9 4 3 7 9 5 5 *